力得文化
Leader Culture

Lead your way. Be your own leader!

力得文化
Leader Culture

Lead your way. Be your own leader!

力得文化
Leader Culture

國貿英語溝通術

劉美慧(Amy)◎著

Master English
Communication for
International Trade

用專業英語打通全世界，
和全世界 談 好 生 意

一次掌握 E-mail 與電話英語的關鍵，輕鬆達陣！
一次修鍊國貿溝通秘技與工作心法，自在致勝！

全面啟動國貿英語溝通力

全書內含7大篇國貿必備實務專業：
基礎國貿知識 商務英文E-mail寫作 詢價議價 訂單
出貨付款 問題處理 商務往來，流程全方位。

共計有48篇E-mail情境、12篇電話實例、61頁心法傳授，
讓你英文跳級、國貿升級、溝通晉級、做人頂級！

每單元並有收錄【單字片語一家親】、【好用句型】、【國貿知識補給站】等...
提升學習力，更快談成好生意。

特別規劃《Dear Amy時間》：跟著Amy一起學習國貿溝通與處理心態的小撇步。

專業國貿人的實用寶典

國貿專業面面俱到 精確掌握國貿流程與重點
國貿實務雙管齊下 讓你寫得精彩、說得流利

強力推薦給各國貿系所、進出口貿易公司的從業人員，以及想有效提升職場英語力的讀者。

作者序 Author

　　我是一個好人！吧？除了在笑看人生之際，偶爾藏不住毒舌會造些口業之外，應該可拿到一枚好人勳章吧！我是個有良心的人，我並沒有要求讓良心成為我的主宰，因為，良心本來就是我的主宰！因為有良心，所以我這本書沒有隨便寫寫（呼～先擦擦汗），也因為如此，所以我待人算有真誠，對事還夠認真。這一股認真，讓我對周遭的人與事常感到好奇，有好奇就不時會有新發現。我會不斷發現人的單一行為，其實是由許多的人生經歷與故事而來，多懂了之後，就多學了些寬容。我也時時驗證事情不論多麼無理都有它的邏輯所在，多循著事情的邏輯，就會多了些冷靜。我期許自己遇事冷靜（其實說得客氣了，我是真的還滿冷靜的），我信奉真理與道理，我要求自己處處留心，我相信生活的快意與滿足來自於看事的態度。我是這麼看待人生，也是這麼對待工作中的人與事，而這本書裡說到的處事與應對，所依循的也就是這樣的態度，希望在你感受、領受之後，能夠找出引領你人生、工作愈臻完美的態度！

編者序 Editor

　　台灣，美麗島國，與他國來往做生意天經地義，對於國貿人才的需求也從沒少過，而與他國做越洋生意，除了專業的國貿理論，長年累積下來的實務經驗更是談成生意的基本門檻，但最重要的，還是英語能力，想想多少信件往來、申請表格都避免不了要使用英文，若能掌握好英語能力，和老外談生意必能無往不利，您說是吧？

　　全書筆觸有趣又專業十足，包含了基礎的國貿知識、英文 E-mail 寫作的訣竅，以及國際貿易上會遇到的各種狀況，舉凡詢價、報價、下訂單、出貨、催貨到索取樣品證明等等，不多不少全都匯集在此了，絕對是有志成為專業國貿人最實用的寶典。

　　國際貿易，和老外談生意，免不了要溝通，說通了，一切都好談，人與人之間的關係也是如此，拿起這本書吧，它會教您國貿知識、也教您如何精通「國貿英語」，更教您人與人之間的溝通技巧，讓您笑談生意，生意談得輕鬆又愉快。

力得文化編輯群

CONTENTS

目次

CONTENTS

目次

CONTENTS

目次

CONTENTS

目次

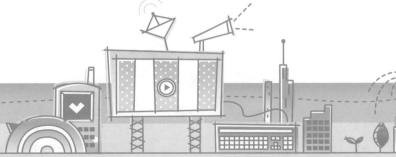

Part 1

蹲好馬步 修煉基本功
1-1 國貿溝通的「溝」要先通一通！

Unit 1　有禮行遍天下
Unit 2　一切都是文化差異
Unit 3　清清楚楚心不慌

國貿英語 溝通術
Master English Communication for International Trade

Unit 01

有禮行遍天下

　　你知道「有禮行遍天下」這句話的真意，在多早以前就有聖哲這樣告誡我們嗎？既然提示是聖哲，那就猜…孔子！答對了，正是孔子是也。在《論語・顏淵》一篇中，孔子說道「君子敬而無失，與人恭而有禮，四海之內皆兄弟也。」意思是君子處事認真謹慎，不會有何失誤、差池，另外，君子對全天下的任何人都能謙恭有敬，以禮相待，因此天下人也會以禮回之，禮尚往來，如此一來，天下人就也就都像是你的兄弟了。

　　在國際貿易的場域裡，我們真的是面對著天下人，而且這個天下的範圍絕對比孔子說的還要大上好幾倍，不過孔子所說的道理卻是亙古不變、放諸四海一樣都合用！如果我們在作業上謹慎不失誤，就可讓國外原廠或客戶不會在收到我們的 e-mail 時覺得頭痛，而若我們提問與回應都能有禮，則國外一般也都會以同樣有禮的態度來對待我們。

　　與國外原廠或客戶溝通時，在沒大代誌的承平日子裡，守禮倒還容易，就像政府推行禮貌運動時要大家將「請、謝謝、對不起」常掛在嘴邊一樣，不難，多說、記得說就是。但若遇到大事，需要跟國外好好談一談，好好釐清問題的責任時，在這個論理的對峙時刻就要更注重遣詞用句了，因為一個不小心，一個失禮，就會讓原先想談理的對方也變得滿身刺，讓雙方的情緒都飆了出來，若到了這個地步，該有的理性對談就不知被壓到哪兒去了！在談論問題時，怎樣才能說寫都能顧及了理，又不失禮呢？而就算是在平時承平時候，又該怎樣才是確保有禮的表達方式呢？就讓我們繼續看下去，來一段有禮遊歷吧！

Part 1 · 1-1 · Unit 1 有禮行遍天下

PART
1

PART
2

PART
3

PART
4

PART
5

PART
6

PART
7

情境・產品有問題，一定是你的問題！

> We've made our experiment by using your antibody products, but the obtained results are very bad! We've used this kind of antibody for many times and we're sure we operated correctly. So the problem must be from your product! Your poor product wasted our samples and time. Tell us what you'll do to compensate our loss.
>
> 我們用了你們的抗體產品來做實驗，但得到的結果很糟！這類的抗體我們已做過好多次了，操作上確定正確，因此，這次的問題一定是出在你們的產品上！你們的產品品質差，浪費了我們的樣品與時間，告訴我們你們要怎麼補償我們的損失。

◎|✗ 論「無禮」、談「有禮」

1. 人是你殺的！

　　法律上有言，嫌犯在未被定罪之前，都是無辜的，因此，像「情境」中說著「The problem must be from your product!」這般不留任何商討空間的直接指控，就算有理在，也確定是失了禮！

　　你有沒有發現抱怨的一方並沒有說明問題是什麼？這種只反應有問題而不細說問題狀況的 e-mail，其實還頗為常見。問題的發生都會有其原由，因此，在告訴對方發生問題時，就該將原委、將過程描述出來，這樣才能讓你的陳述有理字支撐著，也才能讓國外在資訊的量與內容上能盡量跟你同步。待原廠有了充足的資訊，也才能做最有效的判斷，以加快解決問題的速度。而直接的指控等同關了溝通的門，也丟了該有的禮貌與尊重，非但問題沒解決，反而還製造出更多問題。有理時不見得人人皆能有禮，但無理時鐵定無禮！

2. 一定、必定，Must 搞不定

　　Must 扮演的角色是情態助動詞。情態助動詞有其特定的意義，可表現出發話方的態度與意見。Must 是程度最強烈的助動詞，是發話方直接、直白地向收信人提出強硬的要求，説一就是一，要求一定要做！英國語言哲學家格萊斯（Herbert Paul Grice ）在 1975 年提出了「合作原則」（Cooperative Principle），説明對話的目的就是要理解雙方的意願，以有效完成交際、達到溝通，不過，實際上，人際的溝通常會違反合作原則，會有「對話隱含」（conversational implication）的狀況產生，也就是存在著言外之意，而這種間接的表達方式，不見得會讓溝通失敗，反而可表現出言談之間的禮貌，讓交際能夠順利進行。英國語用學家里奇（Geoffrey Neil Leech）在 1983 年為了補充「合作原則」的不足，進而提出了「禮貌原則」（politeness principle），説明在某些情境之下，發話方會為了要遵守禮貌原則，而選擇違反合作原則，亦即交際雙方所做溝通的目的並非只是為了資訊的交流，同時還須顧及彼此的面子與感受，因此，在對話中，常會優先應用禮貌原則。里奇還提到了語用層級的概念，他認為，言談中若是話語的表達愈間接，就能讓聽的人有更多選擇的餘地，這也就是愈有禮的表達方式。里奇舉例如下：

PART
1
PART
2
PART
3
PART
4
PART
5
PART
6
PART
7

		more direct 較直接	less polite 較不禮貌
1	Answer the phone. 接電話。		
2	I want you to answer the phone. 我要你去接電話。		
3	Will you answer the phone? 你會去接電話嗎？		
4	Can you answer the phone? 你可以接電話嗎？		
5	Would you mind answering the phone? 你是否介意接個電話呢？		
6	Could you possibly answer the phone? 你有沒有可能接個電話呢？		
		less direct 較不直接	more polite 較禮貌

「情境」表達裡所用到的 must，在上表中即等同、屬於最直接的層級 1，最無法顧到禮字，而「情境」中最後一句「Tell us what you'll do to compensate our loss.」，劈頭就以祈使句命令對方，也是直接層級第一、禮貌層級最低的表示法。

3. 主觀有餘、客觀不足

　　説話由主觀或客觀兩個不同角度出發，給對方的感受也就大不相同。主觀的表達方式是發話方明白地表示出該觀點是他自己的意見，對事對人發表他個人的評斷，而客觀則是將發球權交給事實，依據客觀事實的陳述來主導評價，沒有發話方自己的意見與情緒涉入其中。當受話方接收到的訊息是客觀事實的情況，沒有或許帶刺的主觀意見，就不會讓聽者不舒服、有想要翻桌子的衝動，這就是發話方有顧到闡述事件與問題該有的有禮與有理態度了。像是在「情境」表達中説的「We're sure we operated correctly.」，主觀態度表達得可清楚了，但字裡行間也就充斥著「我們就是對，你們少來問」的無禮了。

4. 負負不得正

「情境」中先是飆出「very bad」，後來又刺眼地出現「poor」、「wasted」，簡直會讓看的人愈看心裡愈不舒坦，就算原先想好好處理問題，在一路負、負、負的刺激之下，收件人也就無法心平氣和地回話了。要表達問題的嚴重性，可以，但用了具有負面意涵的字詞，確實傳達了嚴重性，但卻失卻了意欲積極合力解決問題的態度與善意。

我們常聽人說到，人要培養正面思考的能力，因為特別是在遇事或遭逢困境時，若我們能以正面的態度來看待，心中自然就會湧出解決問題的企圖，換句話說，在你選擇坦然面對現實的同時，其實也就進入了思考問題、動手解決問題的前哨站。像在「情境」中「the obtained results are very bad」的敘述，只傳達了實驗結果太糟的負面意涵，而這樣的說法就是我們應該要避免的，應要改以正視問題的正面態度來說明，並具體說出實驗到底什麼數值未達標準的這些客觀的陳述。此外，像在「Your poor product wasted our samples and time.」的說法中，「poor」和「wasted」都是充滿負面意涵的字詞，若要贏得對方的同理心，要讓對方以積極的態度來解決問題，則請改採正面表達方式，可說「在這次實驗中，我已投注了時間，也用掉了所蒐集到的樣本，如今發生問題，需要重新蒐集樣本，對我的實驗排程造成很大的影響…」，請記得，有正面的態度，才會有客觀的陳述，也才能夠達成溝通的效度。

最後，就讓我們來看看「情境」案例中的 e-mail 內文，要怎麼寫才能守禮又有禮！

有禮情境・產品有問題，請協助解決！

We've made our experiment by using your antibody, but the obtained Optical Density (OD) value was very high. We have rich experience in using this antibody and we did follow the procedures stated in the operating manual. Therefore, we assume the problem might originate from the product itself. Attached please find the testing data and also our operating procedure details. Please look into the data and help diagnose ASAP where the problem lies.

In addition, we've used up our collected samples in this experiment. Since it's time consuming for re-collecting samples, it will seriously influence our experiment schedule and we'll suffer from the loss. Thus, we might need to discuss a possible compensation with you. If you have any thoughts about this, please let us know. Thanks.

我們使用了你們的抗體來做實驗，但發現測得的光密度值卻非常高。我們在使用這項抗體上很有經驗，操作上也都是照著您操作手冊上的步驟來做，因此，我們認為問題可能來自於產品本身。在此我們將測試資料及操作流程明細貼於附件，請仔細看看這些資料，協助我們找出問題點。

另外，在這次實驗中，我們已用完了所蒐集到的所有樣品，要重新蒐集頗為耗時，因而這勢必將會嚴重影響我們的實驗排程，我們也將蒙受損失，因此，我們將有可能另再與您討論賠償問題，若您對此有任何想法，也請告知，謝謝。

E-mail 寫作禮儀基本功

除了上述在內容表達上顧及禮貌的要點之外，在 e-mail 寫作上也有基本的禮儀一定要顧。請記得，當你寫 e-mail 給國外時，你代表的其實不只是你個人，還代表著雇用你的公司，你就是企業形象的塑造者，你的書寫、言談風格也就代表了企業文化的一部分。因此，就算你本來就是個不拘小節、豪氣干雲的漢子或妹子，寫 e-mail 時，都請盡量要求自己修煉下列這些寫作禮儀基本功吧！

1. 看到人要叫！

你有沒有收過沒有叫你的名，也沒有署他的名的 e-mail？感覺如何呢？在與親近的朋友連絡時，這樣寫當然無妨，但是在商用 e-mail 上，還是要顧點禮儀，因為不是人人都跟你很熟，你並不能完全熟悉每個人的書寫習慣與感覺，所以，若是你自個兒覺得親切，但其實國外原廠或客戶只覺無禮，你是永遠不會知道的。所以，請就遵守著基本的禮儀，一起營造個有禮世界吧！

2. 前謝後謝！

當你要回覆國外原廠或客戶的 e-mail 時，請先謝過對方，像是這類有禮的謝辭：「Thanks for your inquiry.」（謝謝您詢問）、「Thanks for contacting us.」（謝謝您與我們連絡）、「Thanks for your quotation.」（謝謝您的報價）。你一定找得到可以謝謝對方的點，就算對方寫來的是有著濃濃抱怨味兒的 e-mail，你還是能說「Thanks for letting us know the problem.」（謝謝您讓我們知道有這個問題），所以，請記得開始寫 e-mail 時，就來誠心謝一下！

E-mail 寫到末了處的時候，也請記得再謝一下！這樣絕對不是矯情，也絕對很自然，不會像要寫給男、女朋友的分手信那樣難，硬要說著「我已經不再愛你了！我心裡已經有別人了！非常謝謝你！」…這種時候若還保有這種職場上的禮貌，就太令人難堪啦！回到我們的 e-mail 事，若是要對方盡快回覆，寫著寫著就會有這樣的句子出現：「Please reply at your earliest convenience. Thanks.」，自然而然會用謝謝來為 e-mail 劃下句點，得體且得禮。其他有禮的表達方式，像是「Your prompt response will be highly appreciated.」（若您能迅速回覆，我們將深表感激。），或是「We look forward to receiving your soonest reply. Thanks.」（我們期待盡快收到您的回覆，謝謝。），都能讓你禮

貌謝幕，得體退場。

3. 不要隨便吼！

　　若是一篇英文 e-mail 通篇用的都是大寫字母，就代表著是在「shouting」（狂吼）。書寫時用上英文大寫字母有強調語氣的功用，像是說話時大聲說，因此，若是整篇都用大寫字母，那可就真的是從頭吼到尾，不只像是喉嚨會叫破，也讓人看得兩眼渙散，簡直讓禮字不知都退到哪兒去了！還有一點，全用大寫字母寫的文章實在「難看」，閱讀起來會較為困難，因為我們在閱讀英文、在認字時，其實看的並不是只有字母群，還仰賴了單字的形狀，所以，當單字長得全都是大寫樣時，我們就無法藉由習慣、靠著潛意識來認出它到底是個什麼字。

　　若實在要強調，請控制在真正是重點、怕會被忽略的一、兩個單字上，其實也可用斜體、粗體字來表現，以達到強調的效果。所以，寫 e-mail 時請記得不要這吼那吼，不要讓收 e-mail 的人看得眼花，看得眼睛出油，因為這樣就太沒禮貌了！

　　在這本書中的所有 e-mail 範例，一定不魯莽，一定是頭有禮，尾有禮，中段合情合理又合禮，所以，請盡情讀，配著你要寫的 e-mail 內容來微調、微整，盡興利用吧！

國貿英語 溝通術
Master English Communication for International Trade

Unit
02

一切都是文化差異

　　你有沒有碰過這樣的情境？我們自己或是公司的其他人在看到、聽到國外原廠對所提要求的回覆時，常會説出這類直白、流露真性情的話：「國外怎麼那麼笨啊？就這樣改一下也不會？」，或者是「國外腦筋怎麼那麼死啊？都不會轉彎喔？」從這短短兩句話，你有沒有看出什麼重點呢？有啊！「笨」「死」了啊！喔不！重點在於狀況的起源，在於我們要求「改」、要求「轉彎」！當我們自然而然地流露出「山不轉路轉，路不轉人轉」的彈性時，其實在有些國外原廠看來，或許是不合常規、異想天開的要求呢！

情境 · 就改一下品名吧！

> 人客的要求：
>
> 　　We haven't obtained the import permit for our ordered serum products, but we need them urgently. Please confirm whether you could assist by removing the word : "serum" from the label and then arrange the shipment right away.
>
> 　　我們還沒拿到我們所訂血清產品的進口許可證，但我們急著要貨，請確認您是否可給予協助，將標籤上的「血清」字樣刪除，然後馬上安排出貨。

廠商的堅持：

We are NOT able to remove the word "serum" from the product label. It's not a common and good practice! We'll not ship the goods until you get the import permit.

我們不能刪去產品標籤上的「血清」字樣，這不是一般的好作法！等你們取得進口許可之後，我們才會出貨。

　　客戶之所以會提出這樣的要求，為的就是要在沒有進口許可的情況下，還能進口政府的管制產品。進口管制大致上可分為三個層級：

- **bans – where no import is allowed**
  完全禁止進口，全面防堵！
- **quotas – where the volume of goods is restricted**
  可以進口，但可不能需索無度，對數量可是有管控的！
- **surveillance – where the import of goods is monitored with licenses**
  可以進口，但請來申請進口許可，有了許可才放行。

　　在上述的情境之下，要求的一方問來覺得合情合理─合急著要貨的情，合簡單通關的理！要求的人覺得只要國外原廠一個小小的動作，就可快點拿到貨，不至於延誤工作排程，何「便」而不為？有些原廠確實可依要求刪個字，改個詞，給客戶行個方便，但若是碰到覺得這要求不可思議的國外原廠，他們就會明白地用上 no, not, unable 或 can't 這些字眼來回覆，斬釘截鐵地告訴你不行、不能、不依、不做、不會去做改產品標籤這個動作。雙方這種不同的態度，就是文化差異使然了。

○/✗ 先來說說「文化」

文化是什麼呢？讓我們先從文化「culture」這個英文字來瞧瞧！The word 「culture」stems from a Latin root that means the tilling of the soil. 「culture」這個字源自於拉丁文裡的一個字根，意指對土地的耕種。

culture 是由字根 cult＋字尾 ture 而成，拉丁文 cult 的意思是耕種，再加上個名詞字尾 ture，就代表了耕種這事、耕種的結果。可與 culture 拿來相對應的是 nature 這個字，字根 nat 的意思是 born、birth，是出生，再加上名詞字尾 ture 後，則代表了天生自然。nature 是天生自然的狀態，而 culture 則是後天有耕種這項人類文明活動的結果，也就是文化。

那文化是什麼呢？荷蘭學者霍夫斯坦（Hofstede）定義文化為「the collective programming of the mind distinguishing the members of one group or category of people from another」，說文化是一個群體或族群共同、共享之理念與意識的體現，與其他的群體或族群不同。這裡所說的族群，可為以國家、區域、種族、宗教、職業、組織或是性別而分的各個群體。

霍夫斯坦是首開先河研究不同國家間文化差異的學者，他以 IBM 公司 1967～1973 年間來自於四十個國家的員工為樣本，分析不同國家間的文化差異，提出了四項構面，分別是接受權力距離（power distance）差異的程度、防止不確定性（uncertainty avoidance）的程度、個人主義與集體主義（individualism versus collectivism）的程度，以及男性與女性（masculine versus feminality）主要價值的不同。上述情境例子中說到的國外原廠照規矩走，一絲不苟、一字不改的應對與態度，就是防止不確定性的特徵表現。這一點若是反應在組織結構上，公司就會以制定細如牛毛的規章為最高指導原則，若是反應在接單出貨上，就會是有什麼寫什麼，是什麼就怎麼出。所以說，要改標籤？沒的事！

看待文化差異的態度

對於情境中坦白說不的原廠，是不是他們笨呢？這真的無關智力、無關應變能力了。國外原廠這種拒絕配合的態度，其實是信守原則，有所堅持的表現。所以，在我們踢到這樣的鐵板時，請記得，這樣的鐵板不該惡狠狠地嫌它死板，若國外原廠肯鑽個洞讓客戶過，那謝謝配合，謝謝他們的文化與作業規範還可接受這樣的通融之計。若國外硬是不依，那我們就了然於心，點個頭之後，尊重這樣的差異。瞭解了不同文化的差異之後，才能有效溝通，才能在尊重不同文化的態度下，建立起互信的合作基礎。

文化差異之下的合作模式

國際貿易天天都會有新的狀況，進出口貿易的法規規定也常會有所變更，有可能客戶端今天想要國外原廠改標籤，明天碰到客戶專案或驗收單位的要求，又要來請原廠更改效期，後天呢？有可能又要國外寄來產品的空盒，先做形式上的驗收……這些需要孔急的補救措施，並不是常規中的救急辦法，若是第一次發生狀況，在聯繫過、試了水溫之後，已經知道原廠對這類的特殊要求無法配合，那就請謹記在心，在往後商議專案內容與執行要點時，對於該狀況就應該自行尋求他法解決，不要一而再、再而三地拿這類非常規的作法來要求原廠。問問試試不行嗎？行，但通常不會有什麼效果，對我們公司自己的形象建立也不算是件好事！在國外原廠這樣的文化與作業原則之下，一般來說，他們並不會今天拒絕，明天接受，因此，並不建議頻頻刺探原廠的極限。

還有另一個原因，就是我們在行事上也應該要維護公司的誠信形象。所有問題的提問與要求，都是一家公司性格、文化的呈現。若是小動作的要求頻繁得過頭，不免會讓國外原廠對你這家公司產生取巧的印象。雙方要合作，文化或許有差異，調性或許不同，但總是應該愈來愈趨中，要找到一個讓雙方都舒服、都不感威脅的平衡狀態，如此一來，才能懂對方的理，才能從有效的溝通起始點來共同解決問題，合力處理狀況，齊心創造雙方共同且互利的價值！

Unit 03

清清楚楚心不慌

　　你有沒有過這種感覺？在主管或同事交待事情給你，要你發 e-mail 問國外時，常常覺得怎麼他們都講不清楚，讓你要下筆寫 e-mail 時，就算搔頭皮、拔頭髮，還是很難搞得清楚，最後沒辦法了，就索性就給它心虛地發出去，同時也抱著賭一把的心情，就賭賭看說不定國外猜得出來、看得懂哩！

　　若是對到超有責任感、佛心來著的國外廠商或客戶，雖然他們看不太懂你說的是啥、要的是什麼，但是他們會拔根鳳毛，說著如果你的問題是這樣這樣，答案是如何如何，另外再抓根麟角，說著如果你要的是這類的資料，那所有相關的文件就都以附件送上，請你參考。不過，這樣全自動、全方位的原廠或客戶，碰到實屬幸運，實在是少，因為大多數的人都會這樣回你："I don't understand your question. Please clarify."（我不瞭解你所問的問題，請說明。），看到這樣的回覆，可能你會摸摸鼻子，心裡有著賭輸了的消風感覺，然後再去告訴交辦工作的主管或同事 — 這時候注意了！你可能要準備著當個中間人、旁觀者，皮要繃得緊一點了，因為你等一下就得聽著那位主管或同事對著你，數落著國外，說著像是這樣的話：「啊先回就好了嘛！那麼簡單的問題？幹嘛搞那麼複雜！」、「就已經說啦！還要問？！」，等到這些道地的情緒通通倒給你之後，他們才會再跟你說明一次要跟國外回什麼，很有可能的情況是，你還是聽不甚懂，那就要再去寫一次不清不楚的 e-mail，要再一次心慌心虛，再等著看國外能不能回得對，能不能解救你！這樣的生活簡直不是人過的，簡直就是英文業務秘書典型的日子！哈哈！我們做人怎麼能任人擺佈呢？得好好想個法子，拒絕「心」騷擾！要把所有不懂、不合理、不確實的交辦工作通通搞得一清二楚！現在志氣有了！……啊要怎麼做呢？

◎/✕ 內功 ─ 資訊三千，多取幾瓢名詞動詞飲！

　　求諸人之前，當先求諸己！當別人交派工作給我們時，我們聽得霧煞煞，在你想要向外疏通之前，請先做好我們自己的工作，是什麼呢！說穿了，也就是找資料罷了！知識就是力量，若有聽不懂的產品技術問題，請動手找出產品的說明書或是技術文件，從你收過的 e-mail、文件，或是從原廠網站中的產品資訊，先找出關鍵字的段落。這些技術資料大多落落長，所以請睜大眼睛找，或是打字搜尋可能的字詞，找到後，就請好好地整段來回看個兩遍，包你看完之後一定有所得。就算你對技術說明的內容不見得頭尾都懂，但至少你會看到你用得上的名詞與動詞，這時就請用心、用筆、用電腦記下來。名詞部分差不多就是與產品相關的名稱與專業用語，而動詞部分經常是不查不知，查了就知道原來如此的好用動詞。說到這些行話的動詞，讓人感觸還頗深的，因為，每一個行業自己的行話，在能夠畫龍點睛的動詞上可是琳瑯滿目，大家說得道地說得溜，可是其英文常常不是直接翻譯就對得到正確用詞，這可是會讓你寫得心慌，國外看的時候也容易看得意亂！所以，知道行話的名詞與動詞，對你而言就是一件重要的任務了，有了這層知識之後，才能讓你跟國外說一樣的語言，讓你說出來的像人話，說得讓國外一看就懂！

◎/✕ 外攻 ─ 與同仁的溝通攻防戰

　　修完份內的功夫之後，你就需要走出自己的世界，好好面對主管與同事。做什麼呢？請拿出你的邏輯思考能力，好好理解他們要你跟國外溝通的要點是什麼。若你覺得他們的要求在邏輯上難順出個道理，就請馬上問、直接問，因為這些不通的點都是你下筆寫 e-mail 時的痛苦點，會讓你一句寫完不知該怎麼順到下一句，而寫 e-mail 時這種連連看的辛苦，也只有寫的人自己才懂…所以呢？為了善待自己的腦細胞，就請開口問清楚，問了，懂了，則你在專業上的修為就又多添了一分功力，讓你有更厚實的知識背景，可以在下一次「過招」時，更快抓出事情的邏輯點，問出更聰明的問題 ─ 其實，「更聰明的問題」這一點，就是贏得主管或同事尊重你問問題這事的關鍵了。在很多時候，提出問題這件事並不會讓別人認為你什麼都不懂，反而是讓別人知道你懂的有多少！所以請不要害怕問問題，請你想問時，馬上就有禮、有理地問出你的問題，這絕對對你有利，也絕對會讓你在寫作 e-mail 時更有力！

○∣╳ 變身

　　變身？要變什麼身？這不僅僅要你跨越性別的藩籬，更要你消除海陸的阻隔，能夠變男變女變美變醜…喔！變得有點過頭了哩…其實要變的是你的腦袋！除了先前說的文化差異之外，我們要變、要設身處地思考的還有事件本身！提問的人常會覺得問題問得可清楚了，但要回答的人常會覺得問題很模糊，為什麼有這樣的差異呢？原因就在於不同角色的人需要考量的點並不相同。舉例來說，有好幾家競爭廠商都在搶這個客戶的訂單，你是原廠的在台代理商，要跟原廠要個特價，於是你就發了 e-mail，就寫了一句，請國外原廠給你特價，下面呢？下面就沒有了！收到這樣的 e-mail，十家原廠有九家會回問你一堆問題，剩下的那一家之所以沒問問題，是因為你所說的量還不夠大，他們的價格又很硬，所以沒得容你來議價！原廠為什麼會問一堆問題呢？你想想，若是換成你是原廠，你得評估要不要降價？降多少？這時，你會想知道客戶是誰啊？重要嗎？競爭者有誰？競爭者報的價格是多少？要報到多低的價格才搶得到訂單？這些都是合乎邏輯的思路，無關文化，只要你讓自己切換成對方的角色，就知道會問哪些問題，所以，請再切換回原來的你，先把這些問題的答案問出來、找出來，這樣一來，你發給國外的 e-mail 裡，就會有資訊充足的問題分析，原廠只要按個幾下計算機就可給你你要的答案了，不用今天你問，明天他回，明天你再回覆，但你說的還不夠清楚，後天原廠又再問更多的問題…時間、時效都是成事的關鍵，愈早搞定，愈有機會成功搶單下訂。

◎∣╳ 大同世界

　　負責跟國外聯絡的人，一般來說就會是最瞭解國外脾性與邏輯要求的人，肩負著這樣工作的你，若能修煉上述的內功與外攻本事，假以時日，你就會是與國外溝通的頭號人才，而在對工作掌控能力提升之際，你對工作的滿意度也會提高。從總體的角度來看，當你運用修鍊內功得到的知識，配上變身的本事，在你與公司其他人溝通時，其實也就是在跟同事分享你的智識習得。在你提醒、帶領著同事順著國外的邏輯走時，其實也就是將你的內隱知識（tacit knowledge），轉變為外顯知識（explicit knowledge），這就是知識管理上所說的知識外部化（externalization）。接下來，當公司其他人開始將這些從你身上學到的外顯知識，再轉變成他們自己的內隱知識（亦即內部化／internalization），這整段知識流動、轉變的過程：由裡到外、由外再到裡，由個人到團體、由團體到組織，就是所謂的「知識螺旋」（Knowledge Spiral）。這個概念與理論是由日本知識管理學者野中郁次郎（Ikujiro Nonaka）於 1991 年所提出（見於《哈佛商業評論》中〈知識創造公司〉一文）。

　　公司裡的每一個人都有其專精的知識與能力，當內隱與外顯知識在員工之間、在組織內轉換之後，即可有效地擴大個人與組織的知識範圍，創造出組織的知識。在你長期與國外溝通後，你的理解與思考邏輯會漸漸與國外相同，所以當你將所習得的經驗與知識傳給同事之後，同事處理國外事的思路也會漸漸與你相同，這樣一路外展、同化之後，在思辯上就到達了你同、他同、大家同的大同世界了！

　　公司內部同仁在共同處理一個事件時，若能理解你在熟悉國外思維的基礎下所提出事件的問題點，以尊重你、尊重問題的態度來解你的惑，讓你寫出事件交待得清清楚楚的 e-mail，那麼，你就可淡定、心定地處理所有疑難雜症，以你的理、國外的理，順出問題的理，完滿處理！

Part 1

蹲好馬步 修煉基本功
1-2 商務英文 E-mail 寫作

Unit 01 內在美的修為－向簡明靠攏、跟乏味說不！

　　世界上的第一封 e-mail，是 1971 年底在 Bolt Beranek and Newman 工作的雷‧湯姆林森（Ray Tomlinson）所發出，到了 1990 年代，e-mail 這種新溝通媒介迅速地席捲全球。而現在，若是看到國外公司網站上的 "Contact us" 之下只有地址與電話、傳真號碼，居然沒有 e-mail 地址，就不免讓人猜測這家公司規模一定小到不行！不過，有一類的公司網站並不列出 e-mail 地址，那是因為他們要求問題都是由其網站以留言方式發出。

　　從使用 e-mail 的內容型態來看，湯姆林森所寫的第一封 e-mail，是在跟其他技術人員說明如何從網路發出訊息。到了網路發達的時代，e-mail 已是個在國際貿易上再平常不過的聯絡方式，跟以往的郵寄信件、電報、傳真比起來，e-mail 就是快！因此，有人會提出為了要徹底發揮 e-mail 這種溝通方式的功效，就不要太講究格式、不要管什麼文法，寫就對了，快就是好！蓋伊‧川崎（Guy Kawasaki）在《The Guy Kawasaki Computer Curmudgeon》（《蓋伊‧川崎電腦倔老頭》）一書中寫道 "You ask. I answer. You ask. I answer. You're not supposed to watch the sun set, listen to the surf pound the sun-bleached sand, and sip San Miguel beer as Paco dives for abalone while you craft your e-mail."，說著 e-mail 就是那種快問快答、快節奏的溝通工具，不是可讓你為了寫一封 e-mail，還可以跑到海邊去，在當別人去潛水時，你還能在岸上凝望日落、靜聽海濤拍岸，喝著生力啤酒，對著你要寫的 e-mail 在那兒推敲用哪個字比較好。

　　雖然在這種要求立即反應的前提下，確實不用太苛求國貿往來 e-mail 的寫作修辭表現。不過，還是有幾點是寫作、表達 e-mail 內容應當要注意的地方，最高指導原則是要「簡單明瞭」，第二要求是「別太乏味」！該有怎樣的規則與做法，就請見如下的說明了。

能易就不要行難

你可能會說，我也只會一般簡單的字，寫不出難的啊？！不過，你寫出來的 e-mail 裡是不是偶爾會有些怪怪的字塞在裡頭？這些字可能就是你從漢英字典、網路上查出來的那些你其實一點兒也不熟悉的字。從漢英字典查中文字義的單字時，你會有好幾個選擇，有可能這個你沒見過，那個也沒碰過，於是你就索性抓了個字來寫進 e-mail 裡，這樣風險就來了：這個字義其實並不完全適用你要的意思，還有，它很有可能不適用於你 e-mail 內容的語域，例如在你口語化風格的內文裡，突然插進了一個學術、專有名詞，那這封 e-mail 讀來就會很奇怪。

那要怎麼判斷哪些字合適呢？這時請再找找漢英的姐姐 — 英英吧！英英字典都會有單字的英文解釋，在其解釋中所用的字詞一定是比該單字的層級還容易的字，所以你會比較能夠看得懂，那麼你就可用英英解釋查核一下該單字合不合你用，而且，你也可從英英解釋中，知道你要說的詞，可用哪些較為簡單的詞語來說明，可讓你乾脆換句話說，用在 e-mail 寫作裡。舉例來說，你想說明某個故障「原因不明」，在網路查了之後，你找到了這個字 "agnogenic"，但你不確定可不可用，所以改查英英字典，看到它說 "agnogenic in Medicine：Idiopathic"，你就會發現這個字醫學上才用，指疾病原因不明，屬自發性，也就知道了這字絕對不合你用。所以，就請別捨易行難了，要說問題原因不明，就請簡單說 "The cause of the problem is unknown."，就在字裡行間表現出你清楚俐落的個性吧！

再者，如果你想要告訴國外，此計畫在準備與執行期間若有發生什麼問題，就請馬上打電話跟你聯絡，於是你文思泉湧般地寫出這樣長長的句子："If problems are encountered during preparation and execution phases of this project, please refer to me by telephone without any hesitation."，但用在講求簡潔的商用英文上，其實更應該這樣說："If you have any problems on this project, please call me."，如何？前一句多出來的搔頭時間，就可以讓你起身去泡杯咖啡了耶！

能短就不要長，長短中見韻律

在考英文作文時，會有最少字數的限制，若有時沒那麼多想法，倒就想用長一點的片語來灌字數。不過，在商場上寫英文 e-mail，可就沒這種限制了，反而能短就不要長，能簡潔就不要冗長，不要有太長的修飾。例如你要說明 "perform an analysis of"，做了一項分析，其實用上動詞 "analyze" 就可以了，要寫「有鑒於」某某因素，你寫著 "in view of the fact that"，倒不如就寫 "considering"，寫 "with the exception of"，其實就可寫 "except"，既簡短又明快。

再來，除了長長的片語可用隻字來替代之外，長長的句子也請盡量避免，因為句子的長短與可讀性的高低有關，太長的句子會讓對方讀著讀著就恍了神，容易搞不太清楚句子的重點，也會造成讀者的負擔。寫一篇 e-mail，有的句子長一些，有的短一點，句子長度控制在平均 20 字左右，是讓人較容易理解的句長。同時，光是句子長度的變化，也可讓文章有意思些，增加文章的韻律感，而且短的句子還有加強語氣的效果。不過，若需要說明一些較為複雜的狀況，當然也就非長句不可，以交待清楚為上。

多來點主動，偶爾穿插些被動

所謂的主動語態（active voice），亦即句子裡的主詞就是執行動作的人，被動語態（passive voice）即是主詞是接受行為的人事物。主動語態在句子裡寫來比較簡短，讀起來也較清楚，而且也會較有人情味。像是說著 "It is suggested that this operating system be installed."，倒不如將被動轉主動，改寫為 "We suggested to install this operating system."，直接說著我們建議安裝此作業系統。不過，若是你寫著寫著，發現一路下來句子的開頭都是 "We"，當然這時就該來些變化，減少單調與疲乏，換個被動，讓看的人換個方向想一下，活絡活絡。若你要說 "We have received your reply. We'll process your order shortly. We'll ship the goods out for you once the products are ready."，三個 We 的句子連著下來，是不是快讓你也覺得自己無趣到極點，覺得自己自大傲慢了呢？若覺得句子的結構都一樣，主詞都一樣，唸來讓你自己都快打瞌睡了，那就請起而行，改改結構，像上述的三句話，就可改為 "We have received your reply. Your order will be processed shortly. Once the products are ready, we'll ship the goods out for you."

◎/✗ 來點峰迴路轉，增加流暢度

　　在平鋪直敘的 e-mail 寫作中，其實可以用上一些轉折語，透露一些訊息，讓看 e-mail 的人對即將出現的狀況，在心裡有個譜，也可引起看下去的興趣。不過，轉折語也不要過長，最好是簡潔些，而且，剛剛才用過的轉折語，也不要接著再用一次，或是選個極為相似的用，不要白白地讓轉折又變得乏味了。若是寫出這樣的句子 "We have the ability to promote your products in Taiwan. In addition, we can manage to organize seminars and training programs for you. Additionally, we can successfully expand your brand exposure in Taiwan."，裡頭雖有轉折語氣，但都跟 addition 有關，唸起來不太有趣，也還會讓看的人腦子裡「叮」的一聲，誤解你為無趣的人呢！所以，要說明你們有三多，有能力一、推廣國外的產品，二、辦研討會，也辦訓練課程，三、增加國外品牌的曝光度，既然有這麼強的能力可表現，就請加些不同的轉折語，讓語句生動些，可改為 "Not only do we have the ability to promote your products in Taiwan, but we can also manage to organize seminars and training programs for you. More importantly, we can successfully expand your brand exposure in Taiwan."，這樣就將原來有 addition 的兩個轉折語，轉成了 not only… but also 與 more importantly，語氣多了強度，句子的說服力也更能讓人印象深刻！

Unit 02
讓你有型有格－格式裡的內幕

　　不同的 e-mail 系統有不同的介面與格式，但在組成要素上，倒是不管哪個系統，都有必定存在的幾個要素，這些要素你也都已經熟到不行了。在這裡，我們就來看看各個要素裡頭有哪些要點，有哪些是我們應該注意的地方，讓我們跟這些老朋友更熟一些吧！

No.	組成要素	英文名	內容與內涵
1	收件人	To	填入收件人的 e-mail 地址。
2	副本	CC	CC 是 Carbon Copy 的縮寫，Carbon Copy 是複寫副本，填入要知會、亦可看到此封 e-mail 之第三人的 e-mail 地址。
3	密件副本	BCC	BCC 是 Blind Carbon Copy 的縮寫，BCC 的收件人可看到此 e-mail，但又不會讓檯面上的收件人看到此 e-mail 到底還秘密地入了誰的眼簾！
4	主旨	Subject	填入 e-mail 內容的主題或簡要說明，會顯示在收件人的收件匣目錄上。
5	本文	Message Text Area	牛肉就在這裡！將欲傳達的訊息寫在此處，這部分的文字可使用編輯功能，文字要大、要小、要粗、要斜都可以。
6	附件	Attach-ment	由加入附件處將檔案附上，隨 e-mail 傳送給收件人。

收件人 e-mail 地址 — 有前有後

在收件人與副本處，若要發送的對象多於一人，那就也出現了人員先呼後叫的排序狀況。雖然這實在是個小事，但由小處也可看出些禮儀的細緻度。基本上，秉持著我們一向謙恭有禮的態度，e-mail 地址排列的順序自然而然就會是先國外，再國內，而若是發給同一家公司裡多個成員，e-mail 地址的先後順序，就會以位階由高到低來排列。若是不想捲入這些職位高低的複雜排列，那就請以讓人看得出端倪的排列順序來排，例如以英文名字字首的順序。

副本與密件副本 — 公開與不公開之間

E-mail 中所填入 CC 的這位第三人，屬於事件的次要關係人，通常填入的會是你的主管、公司內部其他相關人員，或是國外公司的主管與相關人員等人。當你將主管列為 CC 副本收件人，其實也在告訴對方這件事是有上報到主管、知會到哪些相關的人員，所以，e-mail 的重要性與涉及範圍其實也可從 CC 處得到資訊。國外回覆時，若是選取「全部回覆」，寄件時的所有相關人員也皆可收到回覆。順道一提，Carbon Copy 的用法源自於打字機的時代，當年打字若要打個兩份，打字紙下方就要先放一張 carbon paper，也就是碳粉複寫紙，使勁點兒打字，就可多打出一張複寫副本了。

而密件副本 BCC 的「B」字，代表 blind，意思也就是在有名分的收件人收到信時，對於 e-mail 還另外再寄給了誰，是 blind，是看不到的，所以稱此副本為 Blind Carbon Copy。

主旨 — 先講先贏

還沒打開 e-mail 時，在目錄中會看到的唯二訊息就是寄件人跟主旨了，因此，寫個一目了然、吸睛的主旨就是個重點了。

主旨的要點請寫在主旨的前半段，因顯現在目錄上時，後半段會被裁剪掉。主旨的第一優先要點為你要請國外做的動作，例如 "Please ship our order today"，讓國外清楚地知道這一封 e-mail 就是要來請他們今天安排出貨。

為了提醒注意，為了要求國外儘速處理，常可見到主旨以大寫字 "URGENT" 寫出，緊急時候說緊急，當然天經地義，但若是 "URGENT" 三天兩頭出現，會使得原來看到 "URGENT" 會緊急處理的國外公司，也都不免疲乏且心生反感了。請

注意，"URGENT" 的發送心態有兩種，一種是國外效率不佳，對所收到的 e-mail 沒個回應、怠慢了，所以我們催得心急，但也理直氣壯，但是，若像是我們當天提出個要求，一定要國外當天回覆，加了 "URGENT"，內文處"URGENT"還加粗體，那對國外的收件人來說，可就頗有點刺眼了，所以若要國外緊急當天處理，請更有禮，以請國外幫忙的口吻來書寫 e-mail。

本文 ─ 綜「意」大集合

E-mail 的寫作要有理又有禮，這在本書 Part 1 Unit 1 已有說明了，在這裡我們就再來看看還有其他什麼地方要注意。

1. 先吃好料，再喝湯

首先，在文章寫作上，"inverted-pyramid" 這種「倒金字塔」的方式。這個方式是記者與作家常用的寫作技巧，也可應用在我們 e-mail 的寫作中，將文章中訊息傳達的優先順序排好，最重要的擺前頭，接著是次重要的內容，文章段落愈往下，訊息的重要性也就愈低。這樣安排的原因在於並非每個人都一定會一封 e-mail 從頭看到尾，所以，在這樣速看速回的本質下，最重要的訊息就一定要搶在最前頭，將狀況與案件的人、事、時、地、原因（who, what, when, where, why）一次列明，確保收信的人一定會讀到。在此將倒金字塔的文章結構圖說列出如下：

PART
1

PART
2

PART
3

PART
4

PART
5

PART
6

PART
7

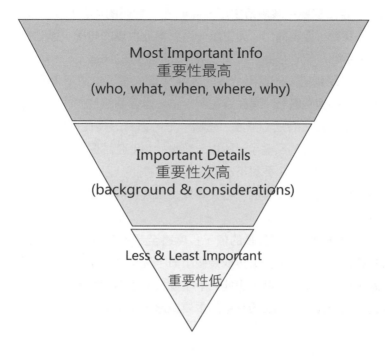

Inverted‐pyramid ／ 倒金字塔

　　有時對重要性高低這種判斷還得好好想一下，若是在討論完 e-mail 主要事件後，另要補充說明即將到來的國定假日放假訊息，想當然爾，它的重要性並不若事件重要，但若國外將此放假期間與所要求提前下單的日程資訊都擺在最後，有的甚至還放到署名之下，那要收件人看到的機率就小多了。這類要求提前下單的訊息雖然簡單，但卻屬於作業安排上的重要訊息，因此，分配在 e-mail 最後頭就不算是個好的處理方式，雖然發信人順道一提了，但反而讓收件人落得一無所知，所以，應當另寫一封通知放假的 e-mail，一歸一，二歸二，不同的重要訊息就寫不同的 e-mail 吧！

2. 要斷就要斷得乾淨，下一個會更好

　　除了由上往下的重點排序寫作方式，在每一個段落裡，同樣地，重點也就必須落在段落起首處，讓人就算只看了一眼，也一定能抓得到該段要說明的主題為何。此外，一個段落裡提到的通常不會只有一個觀點，那既然要兜在一個段落裡，

就請注意這些觀點都要在該段落主題的範疇裡，它們的角色都是要來協助闡述主題。若你寫著寫著，發現寫出了與主題一點兒牽扯也無的句子，那麼就請你快刀斬亂麻，狠心分手吧！讓它移到下一個段落，去找更好的歸宿！

此外，一路說到底、一直不分段的大塊文章，看了也並不會讓人大快人心，只會讓看的人在心裡吶喊著「有完沒完啊！」，也容易讓他們的腦袋決定切換到放空模式，船過水無痕，看過心無感，這樣一來，你寫的長長長長的 e-mail，就常常常常沒有辦法得到好的回應了。所以，寫 e-mail 時請保持清醒，說完了一個觀點、一個層面、一個考量，要說下一個時，就請換下一段來說明。就算整個段落在邏輯上是一脈相承的、是完整的，但只要有文章太長的問題，就一樣請找個換氣的空隙，從中斷開。

當你發狠發狂寫出一封長長的信，長度甚至超過螢幕顯示的一個頁面時，就請你根據內容加註標題，將長文拆成幾個小節，這樣既可讓看信的人抓準何所云，待整篇看完，要回頭找訊息時，也能一下子就找到相關的段落。標題如何下呢？當然別又長到滿滿一行，請挑出段落中的重要概念，簡短就好。

3.「條子」上場

當你有需要在 e-mail 中娓娓道來事情的原委、不同角度的看法，或是執行上的安排順序時，這時你就該以條列式的編排來讓你的表達更清楚，讓看倌順著你一小則一小則的邏輯，不致迷失在大堆頭的訊息之中。尤其是碰到雙方有歧異發生時，像是你要求、提醒過某事，但國外沒照著做，卻又硬說他們有理時，這時就請不要客氣，不要留一手，請將你手邊的資料整理好，依事件進展的日期，一清二楚地將事實條列出來，讓「理」字自己出來發聲，那麼，正義自然就會站在你這邊！……那個，也不是說國外是邪惡的，只是見路有不平，見說法有不正，所以我們請出「條子」幫忙，協助正本溯源囉！

附件 — 跟緊一點！

若有任何表格要填寫，有產品說明書要提供，有合約要簽核，就需要仰賴好用的附件功能來傳送了。在附件部分，要注意的點也是跟整封 e-mail 的宗旨相同：簡單明瞭。若是 e-mail 內文中可以交待的事項，就不要多一道附件的手續，因為你加附件要時間，收信人看附件也要點選、要下載。再來，請注意附件的檔名

要清楚。你的檔案原來都有個檔名，若是中文檔名，或是一個沒那麼切合內容屬性的英文檔名，就請另存新檔，改成一看就知內容主題的檔名。還有，如果是你一次大放送，在 e-mail 中加了好幾份附件，那麼就請你在 e-mail 文中一一說明會有哪些附件，各個附件分別是提供什麼樣不同的資料，而在附件的檔名處，也請加上編號，給個清楚的檔名，讓看信的人邊看 e-mail 內文時，若想先開附件來對照看看，也能不費神地一秒立見！

NOTES

Unit 03

小號協奏曲 -
短小中見功夫！

在 e-mail 文字與格式之外，還有好些縮寫、頭字語、數字，以及標點符號，是我們收發 e-mail 時一定會碰到的短字符號，他們雖然不是那麼重要，雖然不懂、弄糊了、寫錯了也不致於讓你的 e-mail 整個變天，影響成敗，但是，若能將這些「小號」放對位置，打理得整整齊齊，也就會讓你的 e-mail 在整體表現上更無可挑剔！所以，就請一起來看看它們的正確用法，讓我們都能打通這些基礎又簡單的穴點，使整篇 e-mail 暢快不卡卡！

◎/✗ 好說好縮 — 縮寫字

正確使用縮寫確實可省些你打字的時間，也省些對方看詞語的時間，因為常用的縮寫字早已內建在大多數人的腦子裡，讓人瞥一眼就可知其意，想都不用想。縮寫後的字多會加上句點，例如 department 的縮寫為 dept.，另外，像是國名、頭銜、拉丁文的縮寫也都會加上句點，如 U.S.A.、Dr.、etc.。至於包含句點的縮寫字，若是要變為複數時，則要加上「's」，像是若要說會有 15 位醫生參加研討會，就會寫成「Fifteen Dr.'s will attend the seminar.」。接下來，就讓我們來看看有哪些常用的縮寫字，可讓我們快說也快意！。

完整字	縮寫	中文意思
Et alia (and others)	et al.	以及其他人
Et cetera (and so forth)	etc.	等等
Exempli gratia (for example)	e.g.	例如

完整字	縮寫	中文意思	完整字	縮寫	中文意思
approximately	approx.	大約	incorporated	Inc.	公司
building	bldg.	大樓	limited	ltd.	有限的
company	co.	公司	manufactur-ing	mfg.	製造（的）
corporation	corp.	公司	numero (number)	no.	號碼
department	dept.	部門	postscript	P.S.	附註
each	ea	各個	quarter	qtr.	季
government	govt.	政府			

頭頭是道 — 頭字語

　　E-mail 裡有許多的頭字語（acronym）可以應用在常見的片語上，這些字是由數個單字的第一個字母組成，原先是速記人員為了節省打字時間所用，如今在 e-mail 寫作中，仍十分常用，而且這類的「猜謎」也很容易一猜就中，實在讓人有成就感。頭字語在書寫上用的一定是大寫字母，字母跟字母間不用加上句點，結束也不另加句點。常見的頭字語有哪些呢？請見下面這份初階謎題題本與解答吧！

完整片語	頭字語	中文意思
as soon as possible	ASAP	儘快
calendar year	CY	曆年
fiscal year	FY	會計年度
for your information	FYI	讓您知道一下
by the way	BTW	順道一提
in other words	IOW	換句話說
laughing out loud	LOL	大笑

數字

數字在書寫上可用阿拉伯數字，也可用文字來表達，究竟什麼時候該去阿拉伯？什麼時候該擁抱英文文字呢？依循著下列這些原則，就可以讓你寫來不糊塗囉！

數字1 數字在句首時，用文字表示

例 Twelve units of controllers will be sent to you this Friday.

這星期五我們會寄出十二組控制器給您。

數字2 一到九的數字可用文字表示，十或大於十者則用數字表示

例 Our three group members are 15, 25 & 35 years old respectively.

我們這三人小組組員的年紀分別是 15、25、35 歲。

數字3 大的整數數值用文字表示，特定的某數則用數字表示

例 The large shopping mall may lose a million customers a week.

這家大型商場可能一個星期就會流失一百萬名客戶。

例 We have had 6,850 visitors to this exhibition held in World Trade Center.

在世貿中心舉辦的這個展覽，共有 6,850 個人前來參觀。

數字4 百分比用數字表示

例 15 percent、55 percent

◎|✕ 標點符號

當我們與人面對面交談時，對方說話的表情、聲調、音量都可以烘托、加強所傳達出來的訊息內容。而在 e-mail 的寫作上，就是這麼工整的電腦字體，除了在字彙、句型應用上可活絡文章之外，標點符號其實也可補上好些力氣！在這裡，我們來一起看看這些符號有什麼地方是我們要注意的。

句點（period、full stop）

一個句子結束後，就要送它個句號 — 這道理哪算什麼道理啊？但是這可是常見到的錯誤哩！明明有著主詞＋動詞結構的一個完整句子已經結束，但硬是不肯下句點，就還是一路逗下去…其實處處留心真的皆是學問，像這個小小句點，不也就

PART
1

PART
2

PART
3

PART
4

PART
5

PART
6

PART
7

是人生的大道理嗎？若對方情已消逝、心已不在，就請為它正式劃個句點，別「逗」留了啊！

冒號（colon）

當你用到 as follows、the following、below 這些字詞時，也就是冒號要冒出來的時候了，這時候就請讓它出場，為你的陳述內容帶出一連串的條列項目與內容。在正式的稱謂之後，也可使用冒號，例如 "Dear Mr. Collins:"。此外，冒號還常會用在主旨的主標題與副標題中間，像是 "Summer promotion: the hottest and high-quality antibodies!"（夏日促銷：高人氣、高品質的抗體來了！）。

破折號（em dash）

先說說破折號的英文為何叫 "em dash"？請你再唸一次…是的，有唸到"M"這個字母了吧！，它的意思就是說破折號的寬度是以大寫字母中最寬的"M"為基準。破折號所蘊含的表情多，表示強調，也可表示驚訝的氣氛或情緒。要注意的一點是，當前後兩個字詞是同位語時，例如 "We're proud to introduce our new CEO, Dr. Bush!"（我們很榮幸能在此介紹我們的新執行長，布希博士！），new CEO & Bush 這兩個字詞指的是同一人，是同位語，這時候就應該用逗點，不用引來破折號囉。

括弧（parentheses）

之所以要把說的話、寫的字放進括弧裡，表示它就是負責來傳達「附帶」的訊息，也許是要提示一件事，也許是要補充些什麼，也可能是想到了個相關的問題，順道確定一下。但請記得，括弧裡的詞句最好不要比括弧外的句子還長，因為這樣一來，就容易打擾到看信人的思考方向，使得他們在括弧結束時，先前讀到的訊息也已不知道飛到哪兒去了。再者，括弧既然該是附帶訊息該待的所在，就不該將主要訊息放入，像是在寫詢價 e-mail 時，我們會寫出型號、品名、規格，此時，品名應當是主要訊息，也就不應該被打入括弧冷宮裡。

E-mail 距離文謅謅的公文書寫很遠很遠，但是說起 e-mail 該有的內容表達、格式與符號，可也不是三言兩語即可交待得了。這告訴了我們，每件事的背後都有它的道理與考量，若我們能留心用心看待每件事，就會發現它的學問，而有這樣的認真，這些個學問，不就能讓我們成為更好的人嗎？共勉之！

Part 2
價格溝通篇

Unit 1 詢價與報價
Unit 2 議價與調價

國貿英語 溝通術
Master English Communication for International Trade

Unit
01

詢價與報價

單字片語一家親

幾多錢？

定價	list price
報價	quote
有效	valid
到期	expire
運費	freight
貨成本	shipping cost
手續費	handling charge
包裝費	packaging fee
銀行手續費	bank fee
關稅	customs duty
報關費	customs clearance fee

算便宜一點

折扣	discount
折扣率	discount rate
經銷商折扣	distributor discount
一次性折扣	one-time discount
折扣代碼	discount code
數量折扣	quantity discount
大量訂購折扣	bulk discount

怎麼付錢？

付款條件	payment terms
預付	prepayment
信用卡	credit card
電匯	wire transfer

有貨否？

現貨狀況	availability status
有貨	in stock
沒得供貨	unavailable
沒現貨	out of stock
缺貨中	on backorder
出貨準備時間	lead time

品質掛保證

到期日	expiry date
有效期	shelf life
保固	warranty
延伸保固	extended warranty
問題解決程序	troubleshooting procedures

錢外之物

樣品	sample
產品資料表	data sheet
規格	specification
產品說明書	booklet
操作手冊	operating manual

句型

句型 1 要求提供資訊

Please let us know sth.

例 <u>Pleas let us know</u> if you are willing to cooperate with us.

請告知您是否願意與我們合作。

句型 2 提問求解

I would like to know sth.

例 <u>I would like to know</u> whether your product can meet our requirements.

我想要知道您們的產品是否能夠符合我們的需求。

句型 3 表示興趣

We're interested in sth.

例 <u>We are interested in</u> your products and are willing to discuss the possibility of business transaction with your company.

我們對您們的產品有興趣，想要跟 貴公司談談交易的可能性。

句型 4 考慮下訂單

We're currently considering placing an order for sth.

例 <u>We're currently considering placing an order for</u> your automated door systems.

我們正在考慮要下訂單購買您們的自動門系統。

句型 5 還想要問…

One other concern I have is about sth.

例 <u>One other concern I have is about</u> how to keep the shipment at 2-8°C for the transit.

我還想知道的是，您們會如何讓貨的溫度在出貨途中都維持在攝氏 2-8 度。

PART 1
PART 2
PART 3
PART 4
PART 5
PART 6
PART 7

詢價 E-mail | Price Inquiry

詢價 E-mail 範例 1 ·問價、問折扣、問運費、問手續費、要求樣品

To: Oliver Hamilton

Subject: Antibody inquiry

Dear Mr. Hamilton,

We note from your website that you have a wide variety of antibodies and reagents for research use. We're interested in your newly launched immunoassay kit and currently considering placing an order for it. Please let us know its price and availability status. Do you offer any discounts? If yes, please give us the details. Also please advise the estimated freight to Taiwan. If there are any additional fees, e.g. handling fee, please tell us as well. Thanks.

In addition, is that possible for you to supply a free sample for our test? If test results are satisfying, we'll consider placing a large order to you. We look forward to hearing from you. Thanks.

Best wishes,

Olivia Chen
Euphoria Biotechnology, Inc.

單字 ShabuShabu 一小補小補！
☑ variety [vəˋraɪətɪ] (n.) 種種；種類
☑ launch [lɔntʃ] (v.) 推出；發行
☑ euphoria [juˋforɪə] (n.) 幸福；心情愉快

詢價 E-mail 範例 **1** 中文

收件人：奧利佛・漢莫頓

主旨：詢問抗體

漢莫頓先生，您好，

　　我們從 貴公司的網站上，看到了您們有提供研究用的抗體與試劑，而且種類涵蓋很廣。對於您們新推出的免疫試劑組，我們很有興趣，目前正考慮要下單訂購。請告知此產品的價格與現貨狀況，在價格上有任何折扣嗎？若有，還請告知明細。也請告訴我們貨到臺灣的運費預估為何？若還有收取任何其他的費用，像是手續費，也請一併告知，謝謝。

　　此外，請問您可提供免費的樣品供我們測試嗎？若測試後的結果令人滿意，我們將會考慮一次下個大訂單。期待您的回覆，謝謝。

祝好

陳奧莉維雅
幸福生技公司

必 Check 好句

☑ We note from your website that you have…
　我們從您的網站得知……

☑ We look forward to hearing from you.　期待您的回覆。

詢價 E-mail 範例 **2** · 問產品資訊、問價、問現貨、問售後服務

To: Sales Dept.

Subject: Automated Door System inquiry

Dear sir / madam,

I am writing to inquire about your ADS-101 Automated Door System. I have seen the system information on your website and I'm highly interested to learn more about it. I would like to have an idea about its different models, features, and technical specifications. I would also appreciate if you could elaborate on the various models along with their prices, discounts, and availability.

One other concern I have is about the after-sale service. What if the instrument breaks down or develops a fault? In addition, do you offer extended warranty for the instrument? Please reply to me with all relevant information. Thank you very much.

Sincerely,

Michael Gibson
Northwest Instrument Company

單字 ShabuShabu 一小補小補！
- ☑ elaborate [ɪˋlæbəˏret] (v.)　詳細說明
- ☑ break down　故障
- ☑ extend [ɪkˋstɛnd] (v.)　延伸

PART 1

PART 2

PART 3

PART 4

PART 5

PART 6

PART 7

詢價 E-mail 範例 **2** 中文

收件人：業務部門

主旨：詢問自動門系統

> 您好，
>
> 　　我想要詢問貴公司的 ADS-101 自動門系統，我在您們的網站上看到了此設備的資訊，我很有興趣，想要多瞭解些，請告訴我它的各個不同機型與其特色、技術規格，也請您詳細說明一下各不同機型的價格、折扣與供貨狀況。
>
> 　　我們另外還想知道售後服務的相關內容。請問若是機器故障或是出現問題，您會如何處理呢？還有，請問您有提供延伸保固嗎？
>
> 　　還請您告知所有的相關資訊，非常謝謝您。
>
> 謹上
>
> 麥可·吉普森
> 西北儀器公司

Check 好句

☑ I am writing to inquire about…　我想要詢問……

☑ Please reply to me with all relevant information.　請您告知所有的相關資訊。

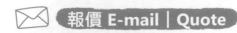

報價 E-mail | Quote

報價 E-mail 範例 1 · 有價、有小折扣、有現貨

Dear Olivia,

Thank you for your inquiry. Please see our quote as below (5% discount off list price is provided):

BA-100 $225/10mg, $675/50mg, $1,035/100mg

The item is in stock and ready to ship once we receive your order. A handling fee of US$ 50 is added to each order to cover packaging. This quotation is valid for 60 days.

The delivery term is F.O.B. Toronto, Canada. Customers are solely responsible for any customs clearance fees required. We ship through FedEx. The shipping cost to Asia is approximately $100. It could be waived from total invoice amount if you can provide us a FedEx account number for shipping. As for payment, please note that prepayment will be required for all new customers.

Please let me know if you have any questions regarding the product and quote. Thanks.

Sincerely,

Oliver Hamilton
Customer Service Department
Atom Chemical Company

單字 ShabuShabu 一小補小補！
☑ solely [`sollɪ] (adv.)　唯一地
☑ approximately [ə`prɑksəmɪtlɪ] (adv.)　大約地
☑ waive [wev] (v.)　撤回、放棄

報價 E-mail 範例 **1** 中文

PART 1

PART 2

PART 3

PART 4

PART 5

PART 6

PART 7

奧莉薇雅，您好，

　　謝謝您發來詢價，請見我方報價如下（定價給 5%折扣）：

　　BA-100 $225/10mg, $675/50mg, $1,035/100mg

　　此品項有現貨，等收到您所發來的訂單後，就可立即出貨。每個訂單都會加計 US$ 50 的手續費，以涵蓋包裝成本。此報價 60 天內有效，出貨條件為 F.O.B. 加拿大多倫多，客戶須自行負擔報關所需的任何費用。我們會透過 FedEx 出貨，出到至亞洲的運費約是$100，若您可提供 FedEx 帳號供我方出貨，則運費將不計入發票總金額。至於付款方面，請注意所有的新客戶皆須事先付款。

　　還請您告知對此產品與報價有無任何問題，謝謝。

謹上

奧利佛・漢莫頓
客服部門
原子化學公司

 好句

☑ Please see our quote as below.　請見我方報價如下。
☑ This quotation is valid for 60 days.　此報價 60 天內有效。

報價 E-mail 範例 2 ‧ 有價、有數量折扣、沒現貨、有樣品

Dear Olivia,

Many thanks for your enquiry. I've attached a quote for the product as requested. This kit is currently in stock.

We do provide quantity discounts on our products. Here are the details:

10 vials or $2,500.00 = 5% off
25 vials or $5,000.00 = 10% off
100 vials or $25,000 = 15% off

We encourage you to make your purchases in these larger quantities to enjoy the discount. Many thanks for writing to us and for your interest in Atom Chemical products.

I have attached the CoA of a batch that we currently have available for your reference and we can supply from it a 50ml sample for you to test.

If you require any additional information, please do not hesitate to contact us. We look forward to your valuable feedback for this quote!

Kind regards,

Oliver Hamilton
Customer Service
Atom Chemical

單字 ShabuShabu 一小補小補！
☑ encourage [ɪn`kɝɪdʒ] (v.)　鼓勵
☑ hesitate [`hɛzə͵tet] (v.)　猶豫
☑ feedback [`fid͵bæk] (n.)　回饋訊息

報價 E-mail 範例 2 中文

PART 1
PART 2
PART 3
PART 4
PART 5
PART 6
PART 7

奧莉薇雅,您好,

非常謝謝您的詢價,在此附上所詢產品的報價單。此產品目前有現貨。

我們對產品確實有提供數量折扣,明細如下:

10 瓶或 $2,500.00 = 5% 折扣
25 瓶或 $5,000.00 = 10%折扣
100 瓶或 $25,000 = 15%折扣

建議您依此數量進行採購,以享折扣優惠。謝謝您與我們連絡,也謝謝您對原子化學產品的興趣。

附上目前我們現貨批次的 CoA 供您參考,我們可自此批次提供 50ml 的樣品供您測試。若您需要任何其他的資訊,請儘管直接與我們連絡,期待收到您對此報價的寶貴意見。

祝好

奧利佛・漢莫頓
客服部門
原子化學公司

必 Check 好句

☑ Many thanks for writing to us and for your interest in Atom Chemical products! 謝謝您與我們連絡,也謝謝您對原子化學產品的興趣。

☑ Please do not hesitate to contact us. 請儘管直接與我們連絡。

報價 E-mail 範例 3 · 有價、沒現貨、提供產品相關資料

Dear Michael,

Thank you for your interest in our ADS-101 Automated Door System. In response to your query, I have attached the data of all models and their detailed specifications for your reference. Below you will find the prices that you requested:

ADS-101-S	Automated Door System	$ 3,000.00
ADS-101-M	Automated Door System	$ 4,000.00
ADS-101-L	Automated Door System	$ 5,000.00

Unfortunately we don't have this system in stock. It will take approximately two weeks to be ready.

In addition to the data provided above, I have also attached the document that contains the information about extended warranty and troubleshooting procedures.

I would like to know your requirements in detail so as to explain to you how our system can help you in that regard. In the meanwhile, if you have any questions or need more clarifications, please do not hesitate to contact me. Looking forward to doing business with you.

Sincerely,

Tim Kershner
Magnificence Electric & Engineering Corp.

單字 ShabuShabu 一小補小補！
☑ unfortunately [ʌnˋfɔrtʃənɪtlɪ] (adv.)　可惜地
☑ clarification [ˌklærəfəˋkeʃən] (n.)　澄清；說明

報價 E-mail 範例 **3** 中文

麥可,您好,

　　謝謝您對我公司 ADS-101 自動門系統表示興趣,對於您所提的問題,我在此予您回覆。茲送上機型與詳細規格資料如附,供您參考,以下為您所要求的價格資訊:

ADS-101-S	自動門系統	$ 3,000.00
ADS-101-M	自動門系統	$ 4,000.00
ADS-101-L	自動門系統	$ 5,000.00

　　可惜此系統現在並無現貨,約兩個星期後才可供貨。
　　除了上述所提供的資料之外,我也附上了有關延伸保固與問題解決程序的資料。

　　我想要更詳細地瞭解您的需求,以能跟您說明我們的系統可如何因應您的需求。同時,若您有任何問題,或是需要更多的說明,請儘管與我們連絡。期待與您合作。

謹上

提姆・克什納
宏偉電機有限公司

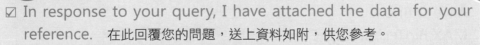

Check 好句

☑ In response to your query, I have attached the data for your reference.　在此回覆您的問題,送上資料如附,供您參考。

☑ In addition to the dataprovided above, I have also attached the document xxx.　除了上述所提供的資料之外,我也附上了 xxx 的文件。

Dear Olivia,

Thank you for your interest in PressMac antibody and for recommending our products to your customers. As you probably know, we do have a distributor in Taiwan, but we are willing to make an exception and sell this product to you. I should mention that we actually have two types of PressMac: SKB-105 which you are already familiar with, and we also have SKB-108 which is a new generation of PressMac. I am enclosing our quote and both antibodies' booklets in this email for you and your customer to take a look at the differences. Let me know which one would be more ideal for your customer.

In order to initiate sale with you, we would need your full shipping and billing addresses, contact telephone number, and if you have FedEx shipping accounts, that number could be used as well. We would insist that a prepayment happen for the first order as well, just as you are a new customer and we are not going through our distributor.

Please let us know if you would want any other information, and we can work together to make this transaction.

Best wishes,

Alice O'Brien
MAX BioTechnology Corp.

單字 ShabuShabu 一小補小補！
- ☑ recommend [ˌrɛkəˋmɛnd] (v.) 推薦
- ☑ ideal [aɪˋdiəl] (adj.) 理想的
- ☑ insist [ɪnˋsɪst] (v.) 堅持

報價 E-mail 範例 4　中文

奧莉薇雅，您好，

　　謝謝您對 PressMac 抗體的興趣，也謝謝您推薦我們的產品給您的客戶。或許您已經知道其實我們已有在台經銷商，不過，我們願意破例一次，銷售此產品給您。我應該要先跟您說明一下，我們的 PressMac 其實有兩款：一款就是您已知道的 SKB-105，我們還有另一款 SKB-108，是 PressMac 新一代的產品。這兩款抗體的產品說明書，我都附在此 e-mail 中了，好讓您及您的客戶瞭解一下兩者間有什麼不同。再請告訴我哪一款產品對您的客戶較合用。

　　為了跟您跟開始合作，還請提供您完整的出貨及發票地址、連絡電話號碼，亦請告知您是否有 FedEx 帳號可供出貨使用。對於首次訂單付款這部分，因為您是我們的新客戶，而我們又沒有透過經銷商交易，所以，我們得要求預付的付款條件。

　　若您還需要任何其他的資訊，再請告訴我們，讓我們一起做成此筆交易。

祝好

艾莉絲‧歐布萊恩
麥克斯生技公司

必 Check 好句

☑ We are willing to make an exception.　我們願意破例一次。
☑ Let me know which one would be more ideal for your customer.　再請告訴我哪一型產品對您的客戶較合用。

報價單格式範例

From:

Quotation

Number	: Q4201
Date	: 9/4/2014
Page	: 1 of 1
Sales order	:
Requisition	:
Mode of delivery	: FXPIC
Terms	: NET30
FOB	: Ann Arbor, MI USA
Quotation expiration	: 9/30/2014

To:

In stock.

Item number	Description	Configuration	Qty	Unit price	Disc %	Disc $	Amount
100079	AUD	50 mg	10.00	$165.00			$1,650.00
100079	AUD	50 mg	20.00	$151.90			$3,038.00

Note:
The quotation provided is for product and services only. Additional charges for transportation, customs, duties, taxes etc... are the responsibility of the customer and added to the invoice upon completion and shipment of the order.

PART
1

PART
2

PART
3

PART
4

PART
5

PART
6

PART
7

報價單格式範例 中文

寄件人：

報價單

號碼	: Q4201
日期	: 9/4/2014
頁數	: 1 of 1
銷售訂單	:
要求	:
出貨方式	: 聯邦快遞優先型
付款條件	: 出貨日後30天付款
FOB離岸價格	: 美國密西根安娜堡
報價單有效期限	: 9/30/2014

收件人：

有現貨.

品項號碼	產品敘述	規格	數量	單價	折扣%	折扣 $	金額
100079	AUD	50 mg	10.00	$165.00			$1,650.00
100079	AUD	50 mg	20.00	$151.90			$3,038.00

備註：
所提供之報價單僅為產品與服務部分的報價，額外費用如運費、報關、關稅、稅負等，皆須由客戶負擔，
待訂單執行完成、出貨完畢時，再另行加入發票中。

☎ 電話對話

電話詢價與報價 範例

人物介紹

Oliver

奧利佛，廠商代表。任職於原子化學公司客服部門，為人乾脆，人客有要求，立馬提供數量折扣。

Amy

艾咪，客戶代表。任職於優越生醫公司，做事迅速確實，是快手，也是好手，值得為她拍拍手。

Oliver: Hello, this is Oliver speaking. How may I help you?

Amy: Hi, this is Amy Liu calling from Excellence Biomedical Company, Taiwan. We're interested in your newest immunoassay kit, DC-200. Could you tell us its price?

Oliver: Sure. How many kits do you need?

Amy: We're thinking of purchasing 1 kit first for evaluation.

Oliver: OK. Its list price is US$ 1,000 per kit.

Amy: Could you offer us a discount?

Oliver: Hmm... We do not give discount for such a small quantity. Would it be possible for you to order more kits in one order?

Amy: Well, what if we order 5 kits in total ?

Oliver: In that case, I can give you 10% discount off. How does that sound?

Amy: It sounds great! Thanks, Oliver.

Oliver: Don't mention it. You're always a good customer to Atom Chemical. When do you think we can expect the order?

Amy: I can place the order to you within 10 minutes!

Oliver: Really? You always amaze me, Amy!

【電話詢價與報價 範例】**中文**

奧利佛：您好，我是奧利佛，有什麼需要我服務的嗎？

艾　咪：您好，我是台灣優越生醫公司的艾咪，我們對您最新的 DC-200 免疫檢驗套組有興趣，能請您報價嗎？

奧利佛：當然，您需要多少組呢？

艾　咪：我們想先買一組評估看看。

奧利佛：好的，每組的定價是美金 1,000 元。

艾　咪：能夠給我們一些折扣嗎？

奧利佛：嗯…對這麼少的數量，我們是沒有提供折扣的。您有可能一次多訂幾組嗎？

艾　咪：嗯，如果我們總共訂個五組呢？

奧利佛：這樣的話，我可以在定價上給您 10%的折扣，這個條件怎麼樣？

艾　咪：很不錯！謝謝了，奧利佛。

奧利佛：不用客氣，您一直都是原子化學的好客戶。請問我們何時會收到您的訂單呢？

艾　咪：我十分鐘之內就可以下訂單給您！

奧利佛：真的嗎？妳總是有辦法讓我感到驚訝，艾咪！

電話英文短句一好說、說好、說得好！

- How may I help you?　有什麼需要我服務的嗎？
- How does that sound?　這個條件怎麼樣？
- It sounds great!　很不錯！
- Don't mention it.　不用客氣。

國貿知識補給站 錢字這條路！

　　價格的說法不只一二，分析起來也是層層掛掛，可有點學問呢！究竟國外報來的是定價，還是折扣後的價格，可得搞個清楚、弄個明白才好。有的國外公司會說「Your cost for #P80 is $300.」，這裡所報的就是買方的產品「cost／成本」或是折扣後的淨價。若說「Your cost will be US list x 0.88.」，這兒明白地寫出「list」，指的就是以美金 list price／定價 x 0.88。那為什麼「list」指的就是定價呢？「list」的意思是表、目錄，而 list price 這種列在表上、價目表裡、目錄中的價格，也就是定價囉！

　　報價若是一次送上兩個價格，像是「$525/100ul (List price); $420/100ul (Transfer price)」，一個是定價，而 Transfer price 就是給買方的價格了！那為何要用 Transfer 這字來表示呢？Transfer 是移轉的意思，所以 Transfer price 指的也就是原廠將貨移轉給買方的交易價格，亦即售價、折扣後淨價是也。

　　再來看看這個報價寫法：「Please kindly note that the unit price of ab192206 is USD 360.」，「unit」是單位、單元的意思，所以 Unit price 就是單價。順道說說「uni-」這個常見字首，它的意思為「單一的」，所以 uniform 說的是單一的型式，應用在服裝上，就是制服；unicorn 說的是單一的 corn，corn=horn／角，因此而成獨角獸，而 unique 這字的意思就是唯一的、獨一無二的意思了。

　　若買方為經銷商，就會看到這種價格說法：「Your distributor price is EUR 250,00 per kit, ex works Bratislava.」，distributor price 就是經銷商價格，價格後頭還接著 ex works Bratislava…這要不要理它呢？不理它好像報的價格也沒出什麼錯啊？不過，說到我們

人要持續學習、要求知，首要條件就是要「好奇」…覺得「好奇」，動手動腦找了答案之後，可能會讓你發現「好神奇」的知識，而這樣的知識累積多了，你就「好神」啦！所以說呢！見到任何我們不懂的詞句，請就多看它幾眼，發揮你的聯想力與想像力，把「不懂」擺平，懂了之後就是你的了！來吧！就讓我們來看看 ex works 所指為何吧！

ex 是「從…」、「在…交貨」的意思，works 指工廠，所以寫了 ex works 這種貿易條件，指的就是工廠交貨價。一般國外廠商所報的價格都是工廠交貨價，不含出貨之後會產生的運費、保險費及稅負等費用，所以，我們平常看國外報來的價格確實都是 ex works 這個條件，確實可以不用理，但現在知道了之後，就提升到懂了所以不用理的層次囉！

說到這裡，就可以來順一次從國外廠商的定價，一直到經銷商報價給客戶這趟價格之旅了：

> **List price 定價**

> **Distributor price 經銷商價格**
> **Discounted price（Disc. price）折扣後價格**
> **Transfer price 國外原廠報給買方的售價**

> **Landed cost 落地成本**
> 貨品到買方公司前的所有成本：
> **產品成本＋運費＋報關費+稅負等成本**

> **End user price 最終使用者／客戶價格**
> （總成本＋利潤）

這些就是代理經銷商會接觸到的幾種相關價格。這些價格在跟國外原廠議價、要求折扣時，可就個個都會跳上檯面來談。若是經銷商要跟原廠要折扣，就得說明成本多高、利潤多薄，所以，此時來個清楚的價格分析，就會是議價成功的關鍵訊息！

國貿實務小小「眉角」

要與國外企業取得聯繫，古早時代要先去翻黃頁、打電報、要跟展覽的主辦單位購買參展廠商的聯絡資料，一切都是來得那麼地用心與用力！而在現在的網站世界裡，動動手指頭就可以「呼風換頁」，實在容易到不行！

而當我們在網路上找到有興趣的廠家之後，怎麼做呢？當然要立馬出動連絡去！連絡資料請在原廠網站右上、左下、右下、左上四處找找「Contact us」／「連絡我們」，點進去之後，就會看到聯絡資訊，包括地址、電話，最重要的就是「E-mail address」，有了它之後，你就可以寫詢價 E-mail 給國外公司了。

不過，有些公司的連絡資料並不會列出 E-mail address，而是請你在網站上留言，請你填寫資料、提出問題。下表即是常見的網站留言表格。這類表格的第一部分會要求你填寫個人資料，第二部分則是重點所在，也就是要你寫出要詢問的問題。這樣的留言表格填來應該是簡單到不行，但反而常會有怎麼樣都無法填對的撞牆狀況出現，怎會如此呢？問題大多出在「State/ Province」及「Postal Code」這兩項資料上，因為我們台灣並沒有「State/ Province」這個州或省的行政分類可填、可選，而在「Postal Code」這一樣，我們填上確實的「Postal Code」，有時又不合乎國外系統的設定，這時候系統就硬是要說我們所填的資料是不正確的。

　　碰到這種說實話無法過關又無從申訴的情況，為求能踏出關係發展的第一步，為了能確實發出留言，就只好委屈一下，先選個系統裡有的 State/ Province。至於 Postal Code，可在網站上查一下該國的郵遞區號位數，然後呢，就請你再勉為其難一次，編寫個系統所接受的郵遞區號吧！等這些小關小卡過了之後，就是大書特書留言內容的時候了，在留言最後，可以特別標註我們的國別，加上個括弧說明，像是「(Country: Taiwan)」，這樣就可確保訊息有清楚傳達了。

PART
1

PART
2

PART
3

PART
4

PART
5

PART
6

PART
7

網站留言格式

Please fill out this form and we will contact you as soon as possible.

* - Required fields

* First Name:	
* Last Name:	
* Organization/Company:	
* Role:	
Industry / Type of Business:	
Address:	
City:	
State/Province:	
Postal Code:	
Country:	
* Email address:	
Phone:	

I would like more information on: ☐ Price Quote

☐ Product Information

☐ Technical Service

Comments:

Submit   Clear

網站留言格式 中文

請填寫下列表格，我們將會盡快與您連絡。

* - 必填項目

*名:	[ ]
*姓:	[ ]
*組織／公司:	[ ]
*職位:	[ ]
產業／營業類型:	[ ]
地址:	[ ]
城市:	[ ]
州／省:	[ ]
郵遞區號:	[ ]
國別:	[ ]
*電子郵件信箱:	[ ]
電話:	[ ]

我想要取得更多資訊： ☐ 產品報價

☐ 產品訊息

☐ 技術服務

留言：[   ]

[送出] [清除]

國貿英語 溝通術
Master English Communication for International Trade

《Dear Amy》時間

Dear Amy，

　　我剛從大學英文系畢業，上星期開始在一家藥品進口貿易公司工作，當英文業務秘書。我沒有什麼國貿背景，面試時，部門主管跟我說進公司後邊做邊學即可。但是，我們部門的資深助理要我跟國外問問題時，說得又快又不解釋，我多問兩句，她就搧著她的兩層假睫毛，不耐煩地跟我說：「反正就是這樣子啦！」我真不喜歡這種感覺，請問我要怎麼做才能很快地備足國貿知識呢？至少要對助理跟我說的事有些概念，這樣我要問也才知道該從哪裡問，該問些什麼。請 Amy 指點一下，謝謝。

不喜歡假睫毛的 Judy

Dear Judy，

　　首先，我要跟妳說明一點，我也不喜歡假睫毛對著我搧風啊！國貿知識的範圍可是包山包海，包貨包錢！美國教育學家杜威（John Dewey）提出「Learning by doing」 這種做中學的方式，確實對非國貿相關科系畢業的人是最實際也最友善的方法，不過，我們會的總是要比別人要求我們的多一些，這樣做起事來才會有成就感，不能永遠處於不太懂的狀態。所以，請好好蒐集一下國貿的相關資料，書本或是網路上整理的皆可，選定一份，認真地從頭看一遍，讓自己好好打個底。在這裡要跟妳說一個可補國貿知識，又可加強商用英文的方法 — 啃英文信用狀！信用狀是國際貿易的一種付款工具，簡單來說，就是開狀銀行開給受益人（賣方）附條件的書面付款承諾。妳可以去找出銀行信用狀的中英文申請表，以及開出來的英文信用狀，一條一條地認真讀過，看出貨須提呈的出貨文件（如 Air Waybills／提單）、看多種價格條件的字首縮寫（如 FOB／離岸價格），而對英文業務秘書來說，更重要的是看正式、正確的國貿知識敘述說明。希望妳儘快把國貿知識和英文這兩個工具磨亮，早日使得上手、使出漂亮，讓自己有自信、有成就感！

PART 1
PART 2
PART 3
PART 4
PART 5
PART 6
PART 7

🔍 🔊 單字片語說分明

• **inquiry** [ɪn`kwaɪrɪ]

n. 詢問 a question intended to get information about someone or something

= query

例 There have already been over 10 inquiries from companies interested!

已經有數家公司有興趣，問了10幾個問題了呢！

其他字義：問題、調查

inquire [ɪn`kwaɪr] Ⅴ

常見搭配詞

price inquiry 詢價　　make inquiries 詢問　　inquire into 調查

- -

• **list price**

n. 定價 a price for a product suggested by the people who make it

例 If no one pays the list price of cars, isn't such pricing meaningless?

如果沒有人以定價來買車，那還列個定價不是沒什麼意義嗎？

Price List 價目表

- -

• **quote** [kwot]

n. 報價 the price that someone says they will charge you for doing a particular piece of work

= quotation

例 You can get an instant quote for all products from our website.

從我們的網站上，你就可以看到我們所有產品的報價了。

- -

• **estimate** [`ɛstə͵met]

v. 估計 to say what you think an amount or value will be, either by guessing or by using available information to calculate it

例 Our estimated shipping date after QC recheck will be around Wednesday. 在重做品管檢測後,我們估計約可於星期三安排出貨。

常見搭配詞:形容詞+名詞 estimate

accurate estimate	realistic estimate	reliable estimate
精確的估計	實際的估計	可靠的估計
conservative estimate	rough estimate	
保守的估計	粗略的估計	

- catalog [`kætəlɔg]

= catalogue(英式拼法)

n. 型錄,目錄 a list or a book containing pictures / information about things you can buy

例 Our full-color catalog contains many new cutting-edge technology products. 我們的全彩型錄列有許多新的尖端科技產品。

縮 cat.

- valid [`vælɪd]

adj. 有效的 legally or officially acceptable

例 The promotion is valid for both domestic (US) and international customers excluding Japan and China. 此促銷方案適用於美國國內與海外客戶,但日本與中國客戶不在此列。

PART 1
PART 2
PART 3
PART 4
PART 5
PART 6
PART 7

・ expiry date

n. 有效期限 the date on which something can no longer be used or is no longer safe to eat

例 The next batch release of this product is estimated to be around the end of July 2015 and this will have an approximate expiry date of June 2017.

這個產品的新批預計將於 2015 年七月底供貨，其效期大約會是到 2017 年六月。

・ packaging [ˈpækɪdʒɪŋ]

n.（商品的）裝箱、包裝、打包 the activity of putting products into containers so that they can be sold in shops

= package、packing

例 Our company supplies environmentally friendly packaging products and solutions and uses recyclable products wherever possible.

我們公司提供環保包裝的產品與解決方案，也儘可能使用回收再生的產品。

・ availability [əˌveləˈbɪlətɪ]

n. 現貨狀況、可得到的物（人）the state of being able to be obtained or used

= availability status

例 Please update us with the availability status of our backorder.

請跟我們更新為出貨訂單的供貨狀況。

・ backorder [bækˈɔrdɚ]

n. 缺貨訂單 an order placed for merchandise that is temporarily out of stock.

例 Please be advised that the product is currently on backorder with an expected release of 1～1.5 weeks.

在此通知您，這一項產品目前缺貨中，預計 1~1.5 個星期後才可供貨。

- customs ['kʌstəmz]

  n. 海關 the place at a port, airport, or border where officials check that the goods that people are bringing into a country are legal, and whether they should pay customs duties

  例 We always declare the true value of our products in the commercial invoice for customs clearance.　我們在商業發票上都是列出產品的真實金額，以辦理清關。

  customs duty 關稅　　　　　　customs declaration 海關申報表
  custom n/adj 習俗；訂製的　　custom-made 訂製的

- wire ['waɪr]

  n. 電匯 to send money directly from one bank to another using an electronic system

  例 All international customers are responsible for duties, taxes, currency exchange, and wire transfer fees.　所有海外客戶皆需自行負擔關稅、稅負、匯率與電匯費用。

  其他字義：n 金屬線　v 給⋯接上電線

- lead time

  n. 出貨準備期 the period of time between the initial phase of a process and the emergence of results, as between the planning and completed manufacture of a product.

  例 If your customer wants 10 kits from the same lot, the lead time would be 2-3 weeks from receipt of order.　如果你的客戶要求 10 組都來自於同一個批次，那麼出貨準備期就是接單後 2～3 個星期的時間。

· specification [ˌspɛsəfəˈkeʃən]

n. 規格；詳細計畫書 an exact measurement or detailed plan about how something is to be made

例 We are confident that we can actually make products that meet your specification requirements.　我們有信心可以確實生產符合您規格要求的產品。

縮 spec.

specific [spɪˈsɪfɪk] adj 特定的　　　specify [spɛsəˌfaɪ] v 指名；具體說明

· procedure [prəˈsidʒɚ]

n. 程序 a way of doing something, especially the correct or usual way

例 We will discuss how to apply some techniques for preventing contamination by incorporating standard procedures into your daily cell culture practice.

我們會討論要如何應用一些技術，並在日常的細胞培養實作時依循標準程序，以避免感染。

· troubleshooting [ˈtrʌbl̩ˈʃutɪŋ]

n. 問題解決；故障排除 the process of finding and repairing problems or faults in something

例 Having trouble with poor resolution? Our Troubleshooting Guide will give you directions to solve common problems.

有解析度不好的問題嗎？　我們的「故障排除指南」裡提供了指示說明方法，可為您解決常見的問題。

國貿英語 溝通術
Master English — Communication for International Trade

Unit 02 議價與調價

📈 單字片語一家親

報價難做

預算	budget
限制	limitation
困難的	difficult
面對（挑戰、困難）	confront
競爭	competition

我們值得更好的價格

勸說	persuasion
關係	relationship
潛力	potential
影響力	influence
忠誠的	loyal
合作	cooperate
努力	strive

要求來著

協商	negotiate
順應	accommodate
共同的	mutual
有利的	beneficial
重新評估	reevaluate
妥協、讓步	compromise
妥協、折衷	meet halfway
比得上（競爭者的價格）	match (competitor's price level)

不調不行

調整	adjustment
通知	notification
定價政策	pricing policy
價格調漲	price increase
成本增加	rising costs
產品成本	production costs
原料成本	raw material costs
包裝成本	packaging costs
人力成本	labor costs
運輸成本	transportation costs

大調或小調

大多數	majority
最少的	minimum
普遍的、廣泛的	widespread
大多數	majority
顯著的	significant
修改	revise
更新	update

調價過渡期

生效	go into effect
生效日	effective date
目前價格	current prices
沿用舊價	honor old pricing
利用	take advantage of
備貨	hold in stock

PART
1

PART
2

PART
3

PART
4

PART
5

PART
6

PART
7

句型

句型 1 反應價格不漂亮

Unfortunately, your price is very high…

例 <u>Unfortunately, your price is very high</u> and our budget does not allow for it.

可惜您的價格太高了，我們的預算無法負擔得了。

句型 2 價格開殺！

Please support by offering xx% off discount…

例 <u>Please support by offering a 30% off discount</u> enabling us to be in a better position to win the tender.

請給予我們 30%的折扣協助，讓我們更有機會贏得這個標案。

句型 3 有難處

Due to production costs, it is rather difficult for us to…

例 <u>Due to production costs, it is rather difficult for us to</u> match competitor's price level.

因為生產成本的考量，要我們的報價比照競爭者的價格水準，實在有困難。

句型 4 更新一下

We have decided to increase prices for sth by xx%.

例 Owing to the high costs, <u>we have decided to increase prices for all products by 5%.</u>

因為成本高，我們已決定將所有產品的價格皆調漲 5%。

句型 5 有效否？

Please let us know how this works for sb./sth.

例 <u>Please let us know how this works for</u> your custom-made instrument.

請告訴我們這對您特製的儀器是否有效？

國貿英語 溝通術
Master English / Communication for International Trade

✉ 議價 E-mail | Price Negotiation

詢價 E-mail 範例 **1** · 價高、預算低、說服點：大咖客戶

Dear Janice,

Thank you for your prompt reply. <u>Unfortunately, your price is very high</u> while our customer's budget is extremely tight. We recommend your product to our customer who used to be a loyal customer to Strength Technology. After our persuasion, the customer is willing to try your product. Now the price is the last thing to build business relationship with this customer.

For your information, the customer is a leading pharmaceutical company in Taiwan. Not only will it have large potential of placing bulk orders, but it also has the power to directly influence lots of other potential customers in the market.

<u>Please support by offering a 40% off discount</u> to the customer, so that price will just be at the same level as Strength's. After the customer places his first order to us, we trust that, with your products of quality and our professional service, we could retain the customer and provide a very positive customer experience so that the customer will want to continue doing business with us and become a loyal customer to your brand. We look forward to hearing from you. Thanks.

Regards,
Richard Huang
Rich Biomedical, Inc.

單字 ShabuShabu 一小補小補！
☑ leading [`lidɪŋ] (adj.)　主要的
☑ influence [`ɪnfluəns] (v.)　影響
☑ retain [rɪ`ten] (v.)　留住；保留

議價 E-mail 範例 **1** 中文

珍妮絲，您好，

　　謝謝您這麼快就回覆我們，不過您報來的價格很高，而我們客戶的預算卻很緊。我們推薦了您的產品給客戶，他們之前是強力科技的忠實客戶，我們說服了他們來試試您的產品，現在與這個客戶建立生意關係的臨門一腳，就只是價格了。

　　讓您知道一下，這個客戶在台灣是製藥公司的龍頭，他們不只有能力訂購大筆的訂單，也有能力直接影響市場上的其他潛在客戶。

　　請您提供協助，給客戶 40％的折扣，這樣的價格才能與強力科技的價格水準相當。等到客戶下了第一份訂單給我們之後，以您高品質的產品、以及我們的專業服務，我們相信可以留住此客戶，讓他們有滿意的體驗經驗，如此一來，客戶就會繼續跟我們購買，成為您品牌的忠實客戶。期待您的回覆，謝謝。

祝好

黃理查
理奇生醫公司

Check 好句

☑ Thank you for your prompt reply.　謝謝您那麼快就回覆我們。
☑ Now the price is the last thing to build business relationship with this customer.
現在與此客戶建立生意關係的臨門一腳就只剩價格了。

Dear Richard,

Thank you for your feedback. Due to production costs, it is rather difficult for us to match Strength's price level. However, in order to compete with Strength and let this important customer switch to buy from us, we can offer you the same level of pricing, a one time 40% discount for this project. Please refer to the quotation number Q2088 when placing your order.

Please fill out the attached form for packaging and concentration instructions. If the customer requires any special processes (such as a custom concentration), this may result in a price change. Typically, it will take 2 weeks from the time of order to prepare the bulk quantity and send to you. We may be able to send more quickly if we already have the raw material in stock. This quotation will expire in one month.

Please let us know how this works for the customer.

Sincerely,

Janice Carrol
Legend Corp.

單字 ShabuShabu 一小補小補！
☑ switch [swɪtʃ] (v.)　轉換；改變
☑ instruction [ɪn`strʌkʃən] (n.)
　指示
☑ typically [`tɪpɪklɪ] (adv.)
　一般地；典型地

回覆議價 E-mail 範例 2 中文

PART
1

PART
2

PART
3

PART
4

PART
5

PART
6

PART
7

理查,您好,

　　謝謝您回饋的訊息。因為生產成本的考量,要我們的報價比照強力科技的價格水準,實在有困難。不過,為了與強力科技競爭,並讓這個重要的客戶轉而跟我們購買,我們答應提供相同的價格,給此專案40%折扣的一次性優惠。下單時請加註報價單號 Q2088。

　　請填寫附件表格,指明包裝與濃度規格,若是客戶有任何特殊處理的要求(如客製濃度),則價格可能會有所不同。我們接單後,一般會需要兩個星期準備像這樣量大的訂單,之後才可出貨。若是原料已有現貨,可能就可以更快出貨。此報價效期為一個月。

　　請告訴我們客戶對此價格的反應。

謹上

珍妮絲・卡羅
傳奇公司

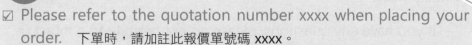

Check 好句

☑ Please refer to the quotation number xxxx when placing your order.　下單時,請加註此報價單號碼 xxxx。

☑ Typically it will take 2 weeks from the time of order to prepare the bulk quantity and send to you.
接單後,一般會需要兩個星期準備像這樣量大的訂單,然後才可出貨。

國貿英語 溝通術
Master English ~ Communication for International Trade

✉ 調價 E-mail │ Price Adjustment

詢價 E-mail 範例 1 · 成本提高、久未調價

Dear Amy,

AB BioTechnologies has been continually striving to meet or exceed customer expectations by providing quality, performing product, and responsive customer service. Due to increasing raw material and packaging costs, we have decided to increase prices for a majority of our immunochemical products by a minimum of 3%. This is the first time in over 10 years we have decided to implement a widespread price adjustment as we have been bearing significantly greater costs compared to the time these products were originally introduced.

Attached please find the PDF file listing the revised immunochemical prices for your review. New prices will go into effect on July 1, 2015 so you have until the end of June to purchase immunochemicals at their current prices.

Please understand we want to do our best to provide spectacular service, but need to follow through with this necessary undertaking. We appreciate your loyalty and hope to continue our business with you by providing satisfying products and services.

If you have any questions, please do not hesitate to contact us.

Regards,

Michael Meissner
AB BioTec-hnologies

單字 ShabuShabu 一小補小補！
☑ responsive [rɪ`spɑnsɪv] (adj.) 回應的
☑ implement [`ɪmpləmənt] (v.) 實施
☑ spectacular [spɛk`tækjələ˞] (adj.) 驚人的；引人注目的

84

調價 E-mail 範例 **1** 中文

PART
1
PART
2
PART
3
PART
4
PART
5
PART
6
PART
7

　　AB 生物科技公司一直以來持續提供高品質、高性能的產品，以及優質的客服回應，以符合或超越客戶的期許。但因為原料成本與包裝成本增加，我們決定調漲大部分免疫化學產品的價格，漲幅會在 3%以上。這次的調整是我們公司在這十年來第一次實行大規模調漲價格，因為相較於產品推出當年，我們現在所負擔的成本已然高出許多。

　　附上免疫化學品項調整後價格的 PDF 檔案，供您過目。新價格將自 2015 年七月一日起生效，因此，於六月底前，您還能夠以免疫化學產品目前的價格來訂購。

　　我們會盡全力提供令人激賞的服務，但也請瞭解我們還是得執行價格調整這個必要的任務。感謝您給予我們的忠誠支持，也希望藉著提供令您滿意的產品與服務，繼續與您合作。

　　若是您有任何問題，請儘管直接與我們連絡。

祝好

麥可梅斯納
AB 生物科技公司

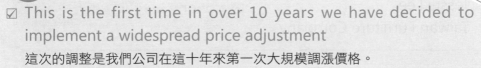

Check 好句

☑ This is the first time in over 10 years we have decided to implement a widespread price adjustment
這次的調整是我們公司在這十年來第一次大規模調漲價格。

☑ New prices will go into effect on July 1, 2015
新價格將自 2015 年七月一日起生效。

調價 E-mail 範例 **2** · 調價前備貨、要求數量折扣

Dear Michael,

　　We have received your notification of price increases. Given the coming pricing adjustment, we will evaluate our market demand and place a bulk order to hold in stock by taking advantage of current prices. In order to evaluate, please provide, the related information of your inventory level for our frequently ordered products, including lot no., quantity and expiration.

　　In addition, please update with us your quantity discount program. We might order more than 100 kits and would like to know what additional discount you could offer.

　　By the way, for our other regular orders that will be placed before June 30, 2015, but shipped after this date, please confirm that you'll still honor the old pricing.

　　We await your reply and confirmation to the above-mentioned points. Thanks.

Best regards,

Jessie Lee
Taiwan Furniture Company

單字 ShabuShabu 一小補小補！
☑ given [`gɪvən] (prep.)　考慮到
☑ program [`progræm] (n.)　方案
☑ honor [`ɑnɚ] (v.)　准允

| 回覆議價 E-mail 範例 2 | 中文 |

PART 1
PART 2
PART 3
PART 4
PART 5
PART 6
PART 7

麥可，您好，

　　我們收到您關於價格調漲的通知了。考慮到價格將要調整了，我們將會評估市場的需求，利用目前的價格，訂購一個量大的訂單，以儲備存貨。為了進行評估，對於我們常訂購的品項，還請您提供其存貨量的資訊，包括批次號碼、數量與效期。

　　此外，也請跟我們更新一下您的數量折扣方案，我們可能會訂超過 100 組，因此想知道一下是否您有額外折扣可提供。

　　順道一提，對於我們其他那些一般性的訂單，若我們在 2015 年六月三十日前下單，但您出貨時已超過此日期，對於這種情形，請確認您仍可遵照舊價提供給我們。

　　我們等您回覆，也等您確認上述幾點問題，謝謝。

祝好

李潔西
臺灣傢俱公司

Check 好句

☑ In addition, please update with us your quantity discount program.　此外，也請跟我們更新一下您的數量折扣方案。

☑ We await your reply and confirmation to the above-mentioned points.　我們等您回覆，也等您確認上述幾點問題。

電話對話

電話議價 範例

人物介紹

Richard

理查，客戶代表，做事抓大方向，夠大器。

Janice

珍妮絲，廠商代表，決斷快，讚美既真且快。

Richard: Hello, Janice. It's Richard from Rich Biomedical.

Janice:　Hi, Richard. Nice to hear from you. How've you been?

Richard: Just fine. I'm calling to discuss with you about your offer yesterday?

Janice:　Well, what did you think of our quote?

Richard: You know Ace Biomed always competes aggressively in the market. Now they just knock down their price to an incredibly low of US$ 2,500.

Janice:　What? I've never heard such a low price!

Richard: Indeed. So could you give us a larger discount?

Janice:　What did you have in mind?

Richard: With an additional 20% off discount, it may be possible for us to win this project.

Janice:　We're willing to support you, but it's way beyond what we can offer. Hmm... I can approve half the requested discount because it's very significant. That's the best we can do! I suggest you carry half the share as well.

Richard: I see. We still appreciate your support! We're prepared to sacrifice our profit to retain this key customer and to secure Legend's market share.

Janice:　Good to hear that! I know you're a reliable and excellent business partner. Legend Corp. is really lucky to have you being our distributor in Taiwan.　Richard: You bet. Hahah…

電話議價 範例 中文

理　查：妳好，珍妮絲，我是理奇生醫公司的理查。

珍妮絲：嗨，理查，很高興你打電話來，你好嗎？

理　查：還可以，我想來跟妳談一談妳昨天的報價。

珍妮絲：好的，你覺得我們的報價如何呢？

理　查：你知道王牌生醫在市場競爭上一直都很積極，現在他們乾脆把價格壓到 US$ 2,500，簡直低得離譜。

珍妮絲：什麼？我從來沒聽過這麼低的價格！

理　查：真的。所以妳有辦法給我們個大一點的折扣嗎？

珍妮絲：那你想要的折扣是多少呢？

理　查：若有額外 20％的折扣，我們就有可能贏得這個案子了。

珍妮絲：我們願意提供協助，但是這個價格真的沒辦法配合，嗯……因為這折扣很大，我可以答應的就只能一半。

理　查：了解，還是很謝謝你的支持！我們有打算犧牲我們的利潤，留住這個重要客戶，這樣也才能保住傳奇公司的市場佔有率。

珍妮絲：很高興聽到你打算這麼做！我就知道你是一個靠得住又讚到不行的生意夥伴，傳奇公司真幸運，能有你來做我們的臺灣經銷商。

理　查：那是當然的，哈哈……

電話英文短句—好說、說好、說得好！

- Nice to hear from you.　很高興你打電話來。
- How've you been?　你好嗎？
- Just fine.　還可以。
- You bet.　那是當然的

國貿知識補給站 有所本、有所議！

　　賣方報了價格，合算、合意的就直接進展到下訂單的程序，但有更多的時候，買方會因購買的量大、賣方報價比競爭廠牌高、價格調漲、預算不夠，甚至就是覺得總要跟賣方要個折扣才快意的這些因素，而有了須跟賣方好好議價的必要。既然要商議價格，那就得在心裡有好幾個數：成本是多少？利潤要抓多少百分比？可接受的最低價是多少？

　　要知道成本是多少，得要知道一下成本的結構（Cost Structure）。廠商會給業務一份價目表，讓業務就照著上面的價格來報給客戶，或是給業務一個基數，讓業務依著原廠價格表或是底價表上的價格，乘上該基數後就得出所定的銷售價格了。就廠商而言，用銷售價格減掉進料、製造、包裝、運輸等成本，即可得出毛利，若要精算淨利，則須納入計算的條目可就多了，像是進貨成本、製造和加工成本、研發費用、包裝費用、運輸費用、出口裝櫃及報關費用、行銷費用、人事及營運成本、稅金、設備資產的折舊等。有了成本與利潤的資訊之後，心中就有個譜、有了把尺，遇到買方來議價時，就可有所本、有概念地來應對了。

　　要不要報高了之後等買方議價、等買方殺呢？在以往的國貿操作中，會有些文化差異的存在，像是亞洲的報價習慣較有可能報高等殺，等著對方殺價，而歐美則較沒有這種行事作風。不過，在現在這個網路世代，原廠的網站就會列出價格資訊，各家競爭廠商的價格水準多已是公開、透明的，所以這種刻意報高等著殺價的案例，就會少上許多。

　　那報價時要不要預留些議價空間呢？開價開太高會讓客戶在第一階段就把你淘汰掉，根本沒機會進到議價階段，所以，報價時一般還是會給個合乎市場行情的價格，留一些合理的議價空間，以能在客戶有

價格要求時，還有商議的可能。

　　那麼，客戶亮出競爭者的價格是不是就是議價的王牌手段呢？也不盡然，市場上常有些新加入的競爭者，來勢洶洶，擺明了就是要搶訂單，所以會報低得出格的價格來「色誘」客戶，而客戶通常抵不住這樣的魅惑，就算合作多年的客戶也可能就這樣隨了人去，徒留你感嘆商場沒有永恆的友誼！原廠對於這樣的客戶，除非有策略性考量，像是要鞏固市場佔有率（market share），要不然其實也並不會每次都隨之起舞，不是都會降到跟競爭者相同的價格，而是會改由品質、從服務上來跟客戶說明他們的優勢，說明他們價格的價值。事實上也確實應該是要如此，因為，決定一個廠商產品或服務的價格，絕對不會是對手的訂價，而是要合乎消費者需求、顧及企業獲利目標的價格，才不會讓員工感覺上整年忙得不亦樂乎，到了結算損益時才發現沒啥利潤，紅利不翼而飛！最後，讓我們來看看一份財務報表，瞧瞧其中與成本相關的條目與英文名稱囉！

Legend Corp.（傳奇公司）		
Operating Income Statement, July 2015（營運損益表—2015 年 7 月）		
Units sold（銷售數量）		1,000
Revenues（收入）		$200,000
Cost of goods sold（銷貨成本）		
Variable manufacturing costs（變動生產成本）	$40,000	
Fixed manufacturing costs（固定生產成本）	80,000	
Total（合計）		120,000
Gross margin（毛利）		80,000
Operating costs（營運成本）		
Variable marketing costs（變動行銷成本）	12,000	
Fixed marketing & administration costs（固定行銷與行政成本）	30,000	
Total operating costs（總營運成本）		42,000
Operating income（營運收入）		$38,000

《Dear Amy》時間

Dear Amy,

　　我們公司是德國藥廠的在台代理商,我是負責跟原廠聯絡的英文業務秘書,幾個月前,我們部門換了個業務主管,有幾次他要我跟原廠要特價時,給我的指令跟內容都是這一句:「去跟原廠要特價」,句點,後頭就沒有任何資訊了,這麼幾次跟原廠要求下來,原廠人員每次都會回說要我提供多一點的訊息,次數一多,我都覺得原廠都說煩了,而且一定覺得我都講不聽、教不會。我要怎麼要求我的主管,要他一次就將資訊給足呢?請幫幫我這夾在中間卻又不知如何施力的小秘書了,謝謝。

想好好教教主管的 Joyce

Dear Joyce,

　　妳這個問題問得實在太實在了,因為很多主管或業務人員要秘書跟國外要求特價時,說的就是這麼簡短,短到讓秘書想多說些、想寫篇文情並茂的議價 e-mail 都苦無素材!原廠對於代理商要求降價,想要知道的不外乎會有這幾點:客戶是誰?有多重要?以後有持續下單的潛力嗎?競爭者有誰?競爭者的報價是多少?代理商希望原廠給的價格或折扣為何?這個案子是一次下單、一次出貨嗎?還是要分批出貨?規格上有沒有特殊的要求?這些資訊給齊了,原廠也才能充分評估情勢。那要怎麼要求主管提供這些資訊呢?就請直接開口問主管,不過,問完後,還是有主管會回說:「先就這樣問出去,再看原廠怎麼回」。這時,小撇步就該用上了…請記得主管或業務都是妳的內部客戶,要讓客戶聽懂我們說的話,就要知道客戶在意的是什麼…若主管在妳問了之後,還不懂得修正做法,那就請跟主管分析,例如:「我們跟原廠要求特價後,原廠都會要問這些問題,若不在第一封 e-mail 中就先說明,原廠就會回問,等我們再補充資訊給原廠,原廠再回我們,這樣 e-mail 一來一往時間就拖長了,就沒辦法馬上回覆客戶了。」這樣妳的主管就會知道問得清楚對他是有影響的囉!

PART
1
PART
2
PART
3
PART
4
PART
5
PART
6
PART
7

🔍 單字片語說分明

• budget [`bʌdʒɪt]

n. 預算 the total sum of money set aside or needed for a purpose

例 With the year-end approaching, our budget is incredibly tight indeed.

年底快要到了，我們的預算實在是非常緊。

常見搭配詞：形容詞／介係詞＋名詞 budget

a tight budget	on/ within budget
預算很緊	在預算內
limited budget	over / under budget
預算有限	超出／低於預算

• limitation [ˌlɪmə`teʃən]

n. 限制 a disadvantage or weak point that makes someone or something less effective

= n. limit

例 The product's design lacks artistic and aesthetic qualities partly due to the operating system limitation.

這個產品的設計欠缺了藝術與美學的質感，部分原因是因為作業系統有其限制。

Limited Company（Ltd. Co.）有限公司

• difficult [`dɪfəˌkəlt]

adj. 困難的 not easy to do, deal with, or understand

例 It's difficult for us to give you an accurate offer until you provide more information about the specifications and quantity you want.

我們很難給你一個精確的報價，除非你多提供些訊息，告訴我們你要的是何規格，要的數量又是多少。

常見搭配詞：動詞＋名詞 difficulty

encounter difficulty	experience difficulty	face difficulty
遇到困難	經歷困難	面臨困難
find difficulty	have difficulty	
發現有困難	有困難	

· confront [kən`frʌnt]

v. 面對；處理 to deal with a difficult situation

例 In talking with many of you, there is an emerging consensus that these are the issues that we must confront together.

從跟你們許多人的談話中，大家慢慢有了一個共識，知道這幾件事是我們必須來共同面對的。

confrontation [ˌkanfrʌn`teʃən] n 對抗；對質

be confront with 面臨

單字小丸子—玩玩字　confront 這個英文字其實很有畫面……「con-」這個字首的意思是「一起、共同、全部」，「front」是「前面」，所以 confront 就是……「踹共！」……是還沒到那麼火爆的氣氛啦！它指的是說當有狀況發生，問題來到我們面前，雖不得已，但也跟它們一起聚了首，這時候我們就該面對、該處理了。

· competition [ˌkampə`tɪʃən]

n. 競爭 the activities of people who are trying to get something that other people also want

例 We have responded to the fierce competition in the market by listening to our customers and offering more effective and efficient customer service.

面對市場激烈的競爭，我們以傾聽客戶的心聲來回應，並提供效能更高、效率更佳的客戶服務。

常見搭配詞：動詞＋名詞 difficulty

cut-throat/ fierce/ stiff/ keen/ intense/ tough competition 激烈的競爭

competitive advantage/ edge 競爭優勢

win/ lose a competition 贏得／輸了競賽

competing products 競爭產品（打對台的產品）

competitive product 有競爭力的產品 （可以打趴對手的產品）

PART 1
PART 2
PART 3
PART 4
PART 5
PART 6
PART 7

• persuasion [pə`sweʒən]

n. 說服；勸說 the process of making someone agree to do something by giving them reasons why they should

例 After much persuasion, the customer finally agreed to place a pilot order to us.

經過我們的努力說服，客戶終於同意簽給我們一個試作訂單。

persuade [pə`swed] Ⓥ 說服；使信服
persuasive [pə`swed] adj 有說服力的；能使人信服的

• significant [sɪg`nɪfəkənt]

adj. 顯著的 very large or noticeable

例 Our company has shown a significant improvement in sales performance in the first quarter.

我們公司第一季的銷售實績有很明顯的改善。

• potential [pə`tɛnʃəl]

adj. 潛在的；可能的 possible or likely in the future

例 The easiest way to identify a potential customer's current supplier is often simply to ask them.

要知道潛在客戶目前的供應商是哪一家，最容易的方法通常就是直接問客戶。

其他字義：n 潛力、可能性

- **negotiate** [nɪˋgoʃɪˌet]

n. 協商；談判 to try to reach an agreement by discussing something in a formal way, especially in a business or political situation

例 You will be required to negotiate contract terms before you are awarded the order.

在拿到這個訂單之前，對方會要求跟你協商合約裡的條款。

- **compromise** [ˋkɑmprəˌmaɪz]

n. 妥協 a way of solving a problem or ending an argument in which both people or groups accept that they cannot have everything they want

例 Please rest assured that we always supply our products at competitive prices with no compromise on quality.

請放心，我們所提供的產品一向都有具競爭力的價格，且在品質上毫不妥協。

其他字義： ☑ 讓步；危及、損害；妥協。

常見搭配詞：

動詞＋a 名詞 compromise	reach/ arrive at/ come to a compromise 達成妥協
動詞 compromise＋名詞	compromise reputation/ safety or security 危及聲譽／安全
a＋名詞 compromise＋名詞 （以名詞修飾名詞）	a compromise solution/ agreement/ settlement 妥協的解決方式／協議／和解

- **mutual** [ˋmjutʃʊəl]

adj. 共同的 belonging to or true of two or more people

例 I would like to meet with you this year to discuss mutual business opportunities between our companies.

今年我想跟您碰個面，討論你我公司共同的商機。

其他字義：相互的

常見搭配詞：形容詞 mutual＋名詞

a mutual benefit	a mutual interest	a mutual friend
一個共同的利益	一個共同的興趣	共同的朋友
mutual respect	mutual trust	mutual support
相互尊重	相互信任	相互支持

· accommodate [ə`kɑmə¸det]

v. 通融 to accept sb's opinions and try to do what they want, especially when their opinions or needs are different from yours

例 Please speak with our Customer Service Department who will make every effort to accommodate your request.

請與我們的客服部門連絡，他們會盡可能地配合您的要求。

· beneficial [¸bɛnə`fɪʃəl]

adj. 有利的 having a good effect

例 I wanted to discuss how a relationship between our two companies can be mutually beneficial.

我想要跟你討論我們兩家公司要如何建立互利的關係。

benefit [`bɛnəfɪt] n 益處；津貼、福利金

beneficiary [¸bɛnə`fɪʃərɪ] n 受益人

Part 3
訂單溝通篇

國貿英語 溝通術
Master English Communication for International Trade

Unit 01

下單與接單

單字片語一家親

訂單要談的有…

型號	Cat. no., Item no., Article no.
品名	product description
條款	term
條件	condition
商品	commodity

網路下單

註冊	register
登入	login
購物車	shopping cart
購物籃	shopping basket
提交	submit
移除	remove
點選	click

請核對確認…

確認通知	acknowledgement
確認	confirmation
處理	process
指派	assign
符合	conform
仔細檢查	look over
準確性	accuracy
如果	in the event of

澄清	clarification
不符	discrepancy
告知	advise

跟總金額有關…

形式發票	proforma invoice
不包含	exclude
幣別	currency
匯兌	exchange
發票開給	bill-to
銀行手續費	bank fee
中間銀行	intermediary bank

跟出貨有關…

出貨	delivery
發貨	dispatch
透過	via
快遞的	express
快遞	courier
立刻地	immediately
優先的	priority
追蹤號碼	tracking no.
延遲	delay
分開地	separately
分批出貨	partial shipment

句型

句型 1 歡喜下單

We are pleased to place an order for sth.

例 <u>We are pleased to place an order for</u> all the following products.

我們很高興要來下訂單給您，訂購下列所有的產品。

句型 2 何時可出

Please let us know when sth can be shipped at the soonest.

例 <u>Please let us know when</u> the commodities ordered in our PO can be shipped at the soonest.

關於我們這張訂單中所訂的貨品，請告知最快何時可以出貨。

句型 3 雀屏中選

We are delighted that you have chosen us as your + n.

例 <u>We are delighted that you have chosen us as your</u> preferred business partner.

很高興您選擇我們做為您首選的生意夥伴。

句型 4 來確認訂單了！

Please see the attached Sales Order for confirmation of your purchase order no. xxx

例 <u>Please see the attached Sales Order for confirmation of your purchase order no. 168.</u>

對於您所訂單號 168 的採購單，請見銷售確認單如附。

句型 5 有問題馬上聯絡！

In the event of any discrepancies, please advise us.....

例 <u>In the event of any discrepancies, please advise us</u> by phone or e-mail right away.

若有發現任何不符之處，請馬上以電話或 e-mail 與我們連絡。

下單 E-mail | Order Placing

下單 E-mail 範例 ・哪些貨、出到哪、何時付錢

Dear Hank,

As per our discussions yesterday, we are pleased to place an order for 100 boxes of Young Hydrating Mask against Catalog No. HM101. The order details, terms, and conditions of the purchase order are as follows. Please confirm the order and let us know when it can be shipped at the soonest.

1. Order details:

Cat. No.	Product Description	Qty.
HM101	Young Hydrating Mask	100 boxes

2. Bill-to and ship-to details:
Address: 33F, No. 333, Ruiguang Rd., Neihu Dist., Taipei City 114, Taiwan
3. Please ship via FedEx International Priority Service.
4. 100% payment will be made on delivery by wire transfer.

Please feel free to contact us for any clarifications or discrepancy in the order contents.

Best regards,

Ann Chen
DRC International, Inc.

單字 ShabuShabu 一小補小補！
☑ pleased [plizd] (adj.)　高興的
☑ hydrate [`haɪdret] (v.)　使成水合物
☑ content [`kɑntɛnt] (n.)　內容

下單 E-mail 範例 中文

漢克，您好，

　　如我們昨天討論所說，我們很高興要來下訂單給您，訂購 100 盒型號為 HM101 的「水漾保濕面膜」，訂單明細及條款在此列出如下，請回覆確認訂單，並請告知最快何時可出貨。

　　1. 訂單明細：

型號	品名	數量
HM101	水漾保濕面膜	100 盒

2. 發票和出貨明細：
地址：臺灣臺北市內湖區瑞光路 333 號 33 樓（114）
3. 請安排聯邦快遞優先服務型方式出貨。
4. 貨款全額將於交貨後以電匯支付。

　　若您發現此訂單內容有任何需説明或任何不相符之處，請儘管直接與我們連絡。

祝好

陳安
迪爾希國際公司

必 Check 好句

☑ As per our discussions yesterday...　如我們昨天討論所說⋯⋯
☑ We are pleased to place an order for...　我們很高興要來下單訂購⋯⋯

國貿英語 溝通術
Master English Communication for International Trade

接單 E-mail | Order Receiving

接單 E-mail 範例 ・確認訂單、有貨、可依要求出貨

Hi Ann,

Thank you for your email and I hope this finds you well. We are delighted that you have chosen us as your supplier for this order.

Please see the attached Sales Order for confirmation of your purchase order received today. Your order has been processed and assigned the Sales Order Number: S109582. Please look over it to ensure accuracy. In the event of any discrepancies, please advise us immediately by return e-mail.

Item # HM101 Young Hydrating Mask is in stock. The order will be shipped via Federal Express by July 10th, 2015. Please note that all International customers are responsible for duties, taxes, currency exchange, and wire transfer fees.

The tracking number and a copy of the invoice will be emailed to you once your order is dispatched. Thank you again for your order. If we can be of any assistance, please do not hesitate to contact us. Have a lovely evening.

Sincerely,

Hank Costner
O&A Cosmetics Group

單字 ShabuShabu 一小補小補！
☑ delighted [dɪˋlaɪtɪd] (adj.) 高興的
☑ assign [əˋsaɪn] (v.) 分派；指定
☑ immediately [ɪˋmidɪɪtlɪ] (adv.) 立刻

接單 E-mail 範例 中文

嗨，安，

　　收信平安，謝謝您發來 e-mail，很高興您選擇我們做為您這筆訂單的供應商。

　　您今天所發來的採購單，我們在此予您確認，請見附件的銷售訂單。我們已開始處理您的訂單，銷售訂單單號編為 S109582，請詳閱，並確定是否無誤。若有發現任何不符之處，還請立即以 e-mail 回覆告知。

　　水漾保濕面膜品項 # HM101 有現貨，將會在 2015 年七月十日透過聯邦快遞出貨。在此提醒您，所有的海外客戶皆需自行負擔關稅、其他稅負、匯兌及電匯手續費。

　　我們一出貨後，就會將追蹤號與發票影本一併 e-mail 給您。再次謝謝您的訂單，若還有需我們協助的地方，請儘管與我們連絡。祝您有個美好的傍晚時分。

謹上

漢克‧科斯納
O&A 化妝品集團

Check 好句

☑ I hope this finds you well.　收信平安。

☑ In the event of any discrepancies, please advise us immediately
　若有發現任何不符之處，請立即通知我們。

網站訂單格式

Basket

| Add Products | Order details | Confirm order | Order complete |

You have 1 item in your order

Product Description	Product Code	Product Cost	Quantity	Remove
Young Hydrating Mask (minimum quantity: 10)	HM101	$ 40	100	Remove

Total cost: $ 4,000 (excluding courier fees)

Address Details

Invoice Address

33F, No. 333, Ruiguang Rd., Neihu Dist., Taipei City 114, Taiwan

Delivery address

33F, No. 333, Ruiguang Rd., Neihu Dist., Taipei City 114, Taiwan

Contact Details

annchen@drcinternational.com.tw

Courier Notes

Purchase Order number

Pro-forma Invoice (Please check this box only if you require a Proforma Invoice)

Accept Terms and Conditions

| Logout | Select more products | Continue |

網站訂單格式 中文

購物籃

加訂產品	訂單明細	確認訂單	訂單完成

您的訂單中有一個品項

品名	產品編碼	產品成本	數量	移除
水漾保濕面膜 (最少數量要求: 10)	HM101	$ 40	100	移除

總成本：美金 4,000 **元**（不含快遞費）

地址明細

發票地址

臺灣臺北市內湖區瑞光路 333 號 33 樓（114）

出貨地址

臺灣臺北市內湖區瑞光路 333 號 33 樓（114）

連絡明細

annchen@drcinternational.com.tw

快遞相關備註

訂購單號碼

形式發票（若需要出具形式發票，請打勾）☐

接受條款 ☐

登出	選取更多產品	繼續

訂貨確認單格式

Sold To: Energy Health Group 35F, No. 335, Ruiguang Rd., Neihu Dist., Taipei City 114, Taiwan	**Order Acknowledgement**
Sales Order number: 116 Order Date: 2-Feb-2015 Ref. Number: P14101755 Payment Term: Net 30	Purchased by: Jade Shih E-mail: jadeshih@energyhealth. com.tw Your Country: Taiwan CURRENCY: EUR
Ship To:	**Bill To:**
Energy Health Group 35F, No. 335, Ruiguang Rd., Neihu Dist., Taipei City 114, Taiwan	Energy Health Group 35F, No. 335, Ruiguang Rd., Neihu Dist., Taipei City 114, Taiwan

Dear Sir / Madam,
Thank you for your order which we confirm as follows:

Please note that items may ship separately.
Please email customer service at order@aquasystems.com
for most accurate shipping date.

ITEM DESCRIPTION	Est. Ship Date	Spec.	Qty.	Unit price
Cat.# SNMW100, Sparkling Natural Mineral Water	10-Feb-2015	750ml	10,000	5
			Tax:	0
			Shipping and handling:	100
			TOTAL AMOUNT:	50,100

Invoice number(s) MUST be included with payment. Please pay the full amount of the invoice without deducting bank fees. Customers are responsible for any correspondent and/or intermediary bank transaction fees.

訂貨確認單格式 中文

買方: 能量健康集團 臺灣臺北市內湖區瑞光路 335 號 35 樓（114）	**Order Acknowledgement**
銷售訂單單號：116 訂單日期： 2015 年 2 月 2 日 查詢號碼： P14101755 付款條件： 淨 30 天	採購人：Jade Shih E-mail：jadeshih@energyhealth.com.tw 國別： 台灣 幣別： 歐元
收貨方明細:	付款方明細:
能量健康集團 臺灣臺北市內湖區瑞光路 335 號 35 樓（114）	能量健康集團 臺灣臺北市內湖區瑞光路 335 號 35 樓（114）

您好， 感謝下單，茲確認如下：	請注意，所訂之品項可能會分開出貨。 若欲查詢最準確的出貨時間，請與我方客服部門連絡，e-mail 地址為 order@aquasystems.com

品項描述	預估出貨日	規格	數量	單價
型號 SNMW100，氣泡天然礦泉水	2015 年 2 月 10 日	750ml	10,000	5
			稅額：	0
			出貨與手續費：	100
			總金額：	50,100

付款時必須加註發票號碼，付款金額須為發票全額，銀行手續費不得從中扣除，客戶需自行負擔任何代理銀行及／或中間銀行的交易手續費。

電話對話

電話下單 範例

人物介紹

Hank

漢克，廠商代表，任職於 O&A 國際公司，辦事有力，照顧客戶有心，客戶關係經營一把罩。

Ann

安，客戶代表，任職於迪爾希國際公司，為人開朗有禮，自然而然建立與廠商的朋友情誼。

Ann: Hello. This is Ann from DRC International, Taiwan. May I speak with Mr. Hank Costner, please?

Hank: Speaking. Hi. Ann, how are you today?

Ann: Not bad! I'm calling to tell you a good news.

Hank: Really? I'm just hoping for good news!

Ann: Hahhah... We've made a decision about this project... We're prepared to place an order to you for Magic Moisturizing Mask.

Hank: Great to hear that! How many do you need?

Ann: We'd like to start with an initial order of 2,000 boxes. If customers respond well, we'll then order a bulk order to you.

Hank: That's the second good news! I will give you a very special discount for your bulk order.

Ann: Awesome! Before I forget, please tell me when the 2,000 boxes can be shipped to us at the soonest.

Hank: Normally, they'll be ready within 7 working days after receipt of orders. For your order, I'll try to speed up and ship within 3 days!

Ann: You're so nice to our company, Brother Hank! Thanks a million!

PART
1

PART
2

PART
3

PART
4

PART
5

PART
6

PART
7

電話下單 範例 中文

安： 你好，我是臺灣迪爾希國際公司的安，麻煩請找漢克·科斯納先生。

漢克：我就是，嗨，安，妳今天好嗎？

安： 還不錯！我打來要告訴你一個好消息。

漢克：真的嗎？我正想聽好消息！

安： 哈哈⋯這個案子我們已經有決定了⋯我們準備要下訂單給你，訂購神奇保濕面膜。

漢克：太棒了！妳要多少數量呢？

安： 第一次訂這個產品，我們想先訂個 2,000 盒，如果客戶反應好，我們就會下個大單給你。

漢克：這是第二個好消息！我會給妳的大單一個很特別的折扣。

安： 讚啦！怕我待會兒忘了，請告訴我這 2,000 盒最快什麼時候可出貨呢？

漢克：一般是 7 個工作天內可出貨，對妳這一張訂單，我會盡量趕在 3 天內出貨。

安： 你對我們公司真好，漢克哥！非常謝謝你啦！

電話英文短句—好說、說好、說得好！

- Not bad!　還不錯！
- Great to hear that.　太棒了！
- Awesome!　讚啦！
- Before I forget, please tell me...　怕我待會兒忘了，請告訴我⋯⋯
- Thanks a million!　非常謝謝的啦！

國貿知識補給站　下單五要！

人要訂目標、訂終身，若要買貨的話，就要訂訂單了！要做到「訂」這個決定、這個動作，我們可就不能隨便了。訂錯麻煩，還要退貨，而訂得不仔細，原廠來的貨、安排的方式就有可能不對、不合意。所以，我們在這就來看看下單這個程序中有哪些 Dos and Don'ts，看有哪些守則可以讓我們訂得完整、訂得漂亮！

1. 下單(Order Placing) 一 要快

評估產品、評估訂購案時仔細為上！當一確定伊人就是你尋尋覓覓的對象後，下聘就要快了！採購案一經確認，請趁熱下單，以免有任何變動發生，尤其是對於原廠發來報價單的特價案件，報價單一般都有效期，過了效期才送訂單過去，不是價格真的已無效，就是你得耗費唇舌拜託原廠了。

2. 產品訊息(Product Information) 一 要全

訂單上要確實寫出要下訂產品的型號、品名、規格、數量。若是賣方有訂單格式，或是買方自己公司裡有制式訂購單，請確實填寫，寫完打完之後呢？請睜大眼睛掃視一次，將數量「2」打成「22」這類的事都不是新鮮事，若是真發生在你小手指打顫、小眼睛脫了窗之際，那麼接下來嚇人的多來貨、多賠錢的事，每次就都很新鮮……因為每次都會像第一次那樣的心痛、怨嘆啊！

3. 日期 (Dates) 一要準

訂單上要清楚列出下單的日期、要求的出貨日期或是到貨日期。下單日期記錄正確才不會在溝通上出問題，它也會關係到銷量統計的期間認定。說到日期，美式日期寫法為「月／日／年」，而英式日期則為「日／月／年」，當然月或日中有超過 12 者，就一定是日了，若數字都小於 12，那就請在原廠來過的資料中找找，確定一下原廠的日期

寫法，這樣就不會出錯了。另外，若是有簽合約、有標案的訂單，關乎罰則的日期會是交貨給採購單位的日期，那就得仔細算回最保險的原廠出貨日，再在訂單上將出貨日跟交貨日都列明，這樣原廠也可配合，不會因資訊不足而造成處理失當。

4. 出貨方式（Mode of Shipment）— 要講

出貨方式有好幾種，有天上飛的（air freight），有海上飄的（sea freight），還有快快飛的快遞公司（express, courier），以及慢慢來的郵局快捷（postal international EMS）。快遞的運送安排還有幾個重點，看是運費要賣方付（freight prepaid）或買方付（freight collect），而有簽約帳號者還要加註寫出。另外，在速度上是要最快的優先處理型（priority service），還是一般經濟型（economy service）？這些都是必要訊息，須指明說清，才不會買方急得一頭熱，賣方卻涼在那裡，整個狀況外！

5. 付款安排（Mode of Payment & Terms of Payment）— 要寫

付款方式（Mode of Payment）有匯款（wire transfer）、支票（check）、信用狀（Letter of Credit）等，若金額不大，刷卡手續費不致於高得嚇人的話，信用卡（credit card）付款通常會是優先選擇的付款方式。另一個有關付款的要點是時間點，也就是會列在付款條件（Terms of Payment）裡的訊息，看原廠要求的是出貨前付款（prepayment, wire transfer in advance）、到貨付款（payment after arrival），或是月結 30 天（net 30 days）等條件。付款安排寫清楚，會記部門也才能好好作業，才不至於在不清不楚中付錯或漏付，讓公司添了糊塗帳或損了信用。

在訂單送出前，請從頭到尾確實再認真看幾眼，要看「對」眼才放行！若確實沒出錯，那這幾眼就值得了，因為這就定下了訂單大業的基礎，若有發現錯誤，那這幾眼就又更值得了，因為就這幾眼，可能就消弭了一場誤會、省了 e-mail 來往勞頓、斷了來貨有錯這些問題呢！

《Dear Amy》時間

Dear Amy，

　　我當英文業務秘書當了十幾年了，以前下單都是把助理打好的訂單傳真給國外原廠，後來用上 e-mail 後，訂單就會以附件發給國外，但這幾年，有些原廠要求「一定」要網路下單，雖然我也常在 pchome、露天拍賣網購下單，但一碰上到國外原廠網站下單，總是這裡不是我要的條件，那裡卡卡的，讓我一碰到要網路下單時，我就整個煩躁了起來！請問 Amy，我要怎樣打破心理障礙，讓我接受並享受新世代的便捷，給它網路下單下到通，下到悠哉自在啊？

有點年紀的美魔女 Jessica

Dear Jessica，

　　看到妳的來信，我忍不住要跟妳說說妳一定懂得的電報時代！這種古意甚濃的外貿商務往來通訊機器，到 1990 年代中期都還有電報在發送呢！現在這個年代，網際網路與電子商務變成了不可或缺的平台，許多廠商的網路就開工接單了。網路下單會卡卡，通常就是發生在這兩個程序上：選出貨方式、選付款方式。上一代常說，「結婚有什麼難？找個人結了就是啊！」這一代會回，「就沒對象怎麼結？」是的，問題就在這裡！妳要的出貨、付款方式沒列在網路訂單的選項中，妳就會卡住！若卡在出貨方式的問題，就請先找找有沒有備註欄可供妳寫特別要求，有的話還好辦，寫清楚便是。若沒有備註欄，又逼著妳選，那只好退而求其次，選個妳公司可接受的方式，或是請示了主管再來下單。而在付款方式上，要網路下單前，請有個心理準備，大部分的網路訂單都只接受信用卡，所以下單前請先查清楚公司對於信用卡下單的金額上限要求為何？有了這個準則之後，再來去網站下單，限額內的就爽快給刷，而在限額外的也可爽快登出網站，換發 e-mail 給原廠，要求另行下單。只要知道處理的方法為何，不管有沒有下成訂單，心裡都會是一片清明澄澈，不會讓妳這個美魔女變成「Lady Gaga」，不會「淚滴」又「卡卡」啦！

🔍 單字片語說分明

- order [`ɔrdə-]

  n. 訂單 a request for a product to be made for you or delivered to you

  例 Please note that any orders received after Thursday will be processed and shipped the following Friday.

  請注意，星期四之後所收到的任何訂單，都只能等到下一週的星期五才能處理，才可出貨。

  常見搭配詞：動詞＋名詞 an order

place an order	cancel an order	process an order
下單	取消訂單	處理訂單

  形容詞／名詞＋名詞 order

  bulk order 大筆訂單　　initial order 首次下單　　purchase order 訂購單

- commodity [kə`mɑdətɪ]

  n. 商品 something that can be bought and sold, especially a basic food product or fuel

  例 The energy commodity market is extremely volatile and driven by geopolitical factors.

  能源商品的市場極為複雜多變，而且會受地緣政治因素所影響。

  Commodity code 商品編碼

  CCC Code (Standard Classification of Commodities of the Republic of China Code) 中華民國商品標準分類號列

- terms [tɜ-m]

  n. （協議上的）條件、條款（用複數） the conditions of a legal, business or financial agreement that the people making it accept

  = conditions, provisions

  例 To ensure that you can fulfill the 30-day payment terms, please pay for these invoices as soon as possible.

為了確保你能履行 30 天的付款條件，請儘速支付這些發票的貨款。

其他字義：期限；關係；術語；方面

常見搭配詞：動詞＋名詞 terms

accept the agreement terms
接受協議條款

agree to/on the lease terms
同意租約條款

negotiate the contract terms
協商合約條款

come to terms (with sb)
達成協議

· condition [kənˋdɪʃən]

n. 先決條件、（尤指協議中的）條件 something that must be true or be done before another thing can happen, especially as part of an agreement, law or contract

例 It's important to read the terms and conditions carefully before signing contracts.

簽合約之前仔細閱讀各項條款與條件，是很重要的一件事。

其他字義：n 情況；狀態。v 影響；使適應；保養（頭髮或皮膚）

常見搭配詞：動詞＋名詞 conditions

meet/ satisfy conditions
符合條件

lay down/ impose/ set out conditions
加諸/制訂條件

· description [dɪˋskrɪpʃən]

n. 敘述 a statement about what someone or something is like

例 It's The product description stated on your purchase order is incorrect. Please revise and re-submit.

您訂單上所列的品名不正確，請修改後重傳。

technical description 技術說明書

job description 工作說明書

常見搭配詞：形容詞＋名詞 description

accurate description	brief description	detailed description
精確敘述	簡要描述	詳細說明

動詞＋名詞 description：give/ provide/ issue a description

提出說明；描繪出來

介係詞＋名詞 description：beyond description 難以形容的

・submit [səb`mɪt]

v. 提交 to formally give something to someone so that they can make a decision about it

例 All the information has been submitted to the manufacturer and it's currently under review.

所有的資訊皆已提交給原廠，現正審查中。

其他字義：屈服；遵守；使服從

・acknowledgement [ək`nɑlɪdʒmənt]

n. 確認通知 a letter telling you that someone has received something you sent them

例 I will process your order shortly and send over an acknowledgement.

我馬上就會開始處理您的訂單，並會再寄出確認通知給您。

其他字義：承認；致謝

・conform [kən`fɔrm]

v. 符合 to be similar in form or type; agree (usually followed by to)

例 All products supplied by Seller shall conform to the specifications listed on this Order.

賣方所供應的所有產品皆須符合訂單中所載明之規格。

其他字義：遵照；順從

- clarification [ˌklærəfəˈkeʃən]

    n. 澄清；解釋 an explanation that makes something cleared and easier to understand

    例 I would need further clarification regarding purchase of use before I sanction any purchase of our cell line products. 在批准任何採購我公司細胞株的案子前，我需要進一步的解釋，說明用途為何。

- discrepancy [dɪˈskrɛpənsɪ]

    n. 不符；不一致；差額 a difference between things that should be the same

    例 Upon review of your account, we have found a discrepancy that we hope you can help clear up.

    檢查了您的帳目之後，我們發現有個地方出現差額，希望您可幫忙澄清一下。

- proforma (pro forma) [pro ˈfɔrmə]

    adj.（拉）形式上 as a matter of formality; denoting a standard document or form, especially an invoice sent in advance of or with goods supplied

    例 Please find the Proforma Invoice attached for the order.

    請見此訂單的形式發票如附。

- currency [ˈkɝ-ənsɪ]

    n. 貨幣 the system of money used in a particular country

    例 Please specify the currency which you will be using to pay for this order.

    請您指明要用何種幣別來支付此筆訂單的貨款。

    其他字義：通行；流行

PART
1

PART
2

PART
3

PART
4

PART
5

PART
6

PART
7

・transaction [træn`zækʃən]

n. 交易 the action or process of buying or selling something

例 Please let me know if you have any questions regarding this transaction.

若您對此交易有何問題，就請告訴我們。

常見搭配詞：形容詞＋名詞 transaction

financial transaction 金融交易

commercial transaction 商業交易

常見搭配詞：動詞＋名詞 transaction

enter into/ engage in a transaction 參與交易

Unit 02 修改與取消訂單

單字片語一家親

求新求變

修改	amend, revise, modify correct, alter
改變	change
調整	adjust
增加	increase, raise
減少	decrease, reduce
取消	cancel
更新	update

念舊

原先的	original
維持	remain
保留	retain

應變

接受	accept
核准	approve
例外	exception
不能接受的	unacceptable
拒絕	refuse
必須遵守的	binding
影響	affect

話說原因

延期	delay, postpone
推遲	defer
受挫、延後	set back
耽擱	hold up
暫停	pause
停產	discontinue
中止	suspend
取消	call off
毀（約）	go back on
迫使	force
需求	demand
意外的	unexpected
突然的	sudden
情況	circumstance
疏忽	oversight
筆誤	clerical error
打字錯誤	typo
誤解	misunderstand
溝通失誤	miscommunication

PART
1

PART
2

PART
3

PART
4

PART
5

PART
6

PART
7

句型

句型 1 可修改否？

We'd like to... if at all possible.

例 We'd like to decrease the quantity of our order if at all possible.

如果可能的話，我們想要減少所訂的數量。

句型 2 條件照舊否？

Although..., we hope you will still agree to...

例 Although now the order is divided into two shipments, we hope you will still agree to offer the same prices.

雖然現在這份訂單要分兩次出貨，我們仍希望您能同意提供相同的價格。

句型 3 破例接受

Considering..., we have decided to make an exception.

例 Considering our long-term and pleasant cooperation, we have decided to make an exception this one time

考量我們長期以來合作愉快，我們決定特別破例一次。

句型 4 改不得

N. + is/are not allowed to be + pp. after the order has been confirmed.

例 The order contents are not allowed to be changed after the order has been confirmed.

訂單一經確認，訂單內容就不得再做更改了。

句型 5 有變化要先說

If + N. + happens again in the future, please keep us well informed of any change in circumstances.

例 If something similar happens again in the future, please keep us well informed of any change in circumstances.

若是往後又有類似的事件發生，請讓我們知道任何的新狀況。

✉ 修改訂單 E-mail | Amending an Order

修改訂單 E-mail 範例 · 客戶合約生變、改數量

Dear George,

We have placed our official order (Purchase Order # A0925) to you and already got your confirmation (Sales Order # Y1202). However, we'd like to make some changes to the order if at all possible. One of our major contracts is postponed to a later date, which forces us to adjust the time scale and quantity of our order. We need to change the quantity of our ordered products as described below:

No.	Product	Original quantity	Amended quantity
1	Extension unit EXU-SI	30	15
2	Extension unit MUL	10	5

As for the discount rate, you gave us 20% off for this order. Although the order contents are changed, we hope you will still agree to let the discount rate stay the same. Please let us have your reply and comments. Thanks.

Best regards,

Pamela Chen
Envision Automation &
Control Corp.

單字 ShabuShabu 一小補小補！
☑ major [`medʒɚ] (adj.) 主要的
☑ extension [ɪk`stɛnʃən] (n.) 延長
☑ envision [ɪn`vɪʒən] (n.) 展望

修改訂單 E-mail 範例 中文

喬治，您好，

　　我們已跟您下了一份正式訂單（訂購單單號 A0925），您也已回覆確認（銷售訂單單號 Y1202），但我們現在得修改一下這份訂單。我們有一份重要合約的執行時程被往後延了，導致我們得調整現階段的時間排程與所需數量。我們需要修改我們所訂產品的數量，明細如下所述：

項次	產品	原訂數量	修改後數量
1	Extension unit EXU-SI	30	15
2	Extension unit MUL	10	5

　　至於折扣的部分，此訂單您給了我們八折的折扣，雖然現在訂單內容有所更動，但我們希望您仍能給我們相同的折扣。還請您回覆您的意見了，謝謝。

祝好

陳潘蜜拉
展望自動控制公司

Check 好句

☑ One of our major contracts is postponed to a later date.
　我們有一份重要合約的執行時程被往後延了。

☑ We need to change the quantity of our ordered products as described below.　我們需要修改我們所訂產品的數量，明細如下所述。

回覆修改訂單 E-mail 範例・破例接受修改、調整所給折扣

Hi Pamela,

We're sorry to learn that your contract was postponed. Hope your business will be back to normal very soon. Normally, the quantity of a Purchase Order is not allowed to be decreased after the order has been confirmed. However, considering you're a loyal customer to Beacon, we have decided to make an exception, just this once, and accept this order amendment.

Speaking of the discount rate; however, we're afraid that we couldn't honor the same rate due to now the order value has dropped to below US$ 10,000. According to our discount policy, you'll get 10% off list price.

Attached please find the revised Sales Order. Please review the list of items and pricing on the order for accuracy. If you have any questions, please don't hesitate to contact us.

Best regards,

George Kellner
Beacon Automation

> 單字 ShabuShabu 一小補小補！
> ☑ accuracy [ˈækjərəsɪ] (n.)　準確性
> ☑ beacon [ˈbikn] (n.)　燈塔；信號

回覆修改訂單 E-mail 範例 中文

潘蜜拉,您好,

很遺憾聽到您的一個合約得延期執行了,希望您的生意快快恢復正常。通常在訂單確認後,我們是不接受訂單減量修改的,不過,因為您是必肯的忠實客戶,我們這次特別破例,接受您修改訂單的要求。

但是說到折扣這部分,恐怕我們無法給您同樣的折扣,因為現在的訂單金額已低於美金一萬元,根據我們公司的折扣政策,您可拿到的折扣是定價的九折。

附上修改的銷售確認單,請檢查所列品項與價格是否都正確無誤。若您有任何問題,請儘管與我們連絡。

祝好

喬治·柯爾納
必肯自動化

Check 好句

☑ Hope your business will be back to normal very soon.
希望您的生意快快恢復正常。

☑ Speaking of the discount rate; however, we're afraid that we couldn't honor the same rate.
但是說到折扣這部分,恐怕我們無法給您同樣的折扣。

取消訂單 E-mail | Canceling an Order

取消訂單 E-mail 範例 ‧遲未出貨、取消訂單、要求往後通知最新狀況

Dear Joe,

Our Purchase Order # SC040601 was placed to you on April 15th. According to the information we got from you previously, the products would be shipped within 5 working days after receipt of our order. But till now, it's been over 2 weeks and we still haven't got any news from you about the shipment.

This delay has adversely affected our production arrangement. We couldn't wait any longer and decide to cancel this order. Please confirm the cancellation.

If such a delay happens again in the future, please keep us well informed of any change in circumstances so that we could reschedule and react better.

Best regards,

Lily Lee
Forever Pharmaceutical Co.

單字 ShabuShabu 一小補小補！
- ☑ adversely [æd`vɝslɪ] (adv.) 不利地
- ☑ reschedule [ri`skɛdʒʊ] (v.) 重新安排
- ☑ react [rɪ`ækt] (v.) 反應

取消訂單 E-mail 範例 中文

喬，您好，

　　我們在四月十五日下了訂購單給您，單號 SC040601，根據先前您所告知的訊息，訂單將於接單後五個工作天出貨，但現在已過了兩個星期，我們仍未收到任何出貨的消息。

　　此出貨延誤已影響了我們的生產排程，我們無法再等，所以我們決定取消訂單，請確認。

　　若是往後又有這種出貨延誤發生，請讓我們知道任何的新狀況，好讓我們能妥善地重排計畫，以因應變化。

祝好

莉莉‧李
永遠製藥公司

Check 好句

☑ The products would be shipped within 5 working days after receipt of order.　訂單將於接單後五個工作天出貨。

☑ We couldn't wait any longer and decide to cancel this order. 我們無法再等了，決定要取消訂單。

回覆取消訂單 E-mail 範例・抱歉、快有貨了、有替代產品

Dear Lily,

I am very sorry for the inconvenience that we have brought to you. After I checked with our lab, the order can be shipped to you this Friday once I get your further confirmation. Will this be OK? If yes, we'll prepare the shipment. If not, then we'll confirm the cancellation.

For your information, we just received the raw materials needed for your order and the production is about to be finished any time this week. Please accept our apologies for not keeping you updated with the latest news.

In addition, we can offer on an alternative kit to your ordered product. Please see the attached kit booklet for details. If it suits your application and you're willing to change to order this kit, we can ship it out for you today. Please let us know your decision. Thanks.

Best regards,

Joe Hassan
Beta Diagnostics, Inc.

單字 ShabuShabu 一小補小補！
☑ alternative [ɔl`tɝnətɪv] (adj.)　替代的
☑ application [͵æplə`keʃən] (n.)　應用
☑ diagnostics [͵daɪəg`nɑstɪks] (n.)　診斷法

回覆取消訂單 **E-mail** 範例 中文

PART
1

PART
2

PART
3

PART
4

PART
5

PART
6

PART
7

莉莉，您好，

　　很抱歉為您帶來不便，我有跟我們的實驗室查詢過，您的訂單可在這個星期五出貨，等您跟我們再次確認後就可以安排，請問這樣可否？若可接受，我們就會準備出貨事宜，若不接受，我們就會跟您確認取消訂貨。

　　讓您知道一下，我們剛收到生產您訂單所需要的原料，產品在這星期內差不多就可生產出來了。抱歉沒有跟您更新最新的狀況。

　　此外，對於您所訂的產品，我們還有提供替代的一組產品，詳細資料請見附件的產品說明書。如果這組產品符合您應用所需，如果您願意改訂，那我們今天就可為您安排出貨。請告訴我們您的決定，謝謝。

祝好

喬·哈森
貝塔醫療診斷公司

Check 好句

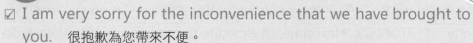

☑ I am very sorry for the inconvenience that we have brought to you.　很抱歉為您帶來不便。

☑ Please accept our apologies for not keeping you updated with the latest news.　抱歉沒有跟您更新最新的狀況。

☎ **電話對話**

電話修改訂單 範例

人物介紹

Elizabeth

伊莉莎白，業務人員，個性大剌剌，有時粗心，但認錯快，不推諉責任，不亂找藉口！

Betty

貝蒂，原廠客服人員，做事俐落，具幽默感，效率佳、笑果足！

Elizabeth: Hello, Betty? This is Elizabeth from Foremost Technology.

Betty: Hi, Elizabeth. What a coincidence! I'm processing your order right now!

Elizabeth: Oh! I should have called earlier! I'm calling to let you know that I have to change our order.

Betty: What happened?

Elizabeth: I just found that I had a typo in the order. I wanted to order "2" kits but wrongly typed "22"!

Betty: That's too bad! I was just happy for you that your sales volume grows significantly!

Elizabeth: Yeah! I also hoped our order was really for 22 kits! Could you agree to change our order quantity?

Betty: That should be alright. We haven't shipped your order out yet.

Elizabeth: What a relief! Sorry indeed for bringing you inconvenience!

Betty: That's OK. Next time, I'll expedite to process and ship your order. I trust that you're absolutely capable of selling out that large quantity very soon!

Elizabeth: Hehe... I'll double check all details before e-mailing orders to you!

PART
1

PART
2

PART
3

PART
4

PART
5

PART
6

PART
7

電話修改訂單 範例 中文

伊莉莎白：妳好，是貝蒂嗎？我是首要科技的伊莉莎白。

貝　　蒂：嗨！伊莉莎白，妳看看有多巧，我正在處理妳們的訂單呢！

伊莉莎白：喔！那我應該再早一點打電話的！我打來就是要告訴妳我得修改我們的訂單。

貝　　蒂：怎麼了嗎？

伊莉莎白：我剛剛發現我訂單打錯字了，我要打「2」組，卻打成「22」組！

貝　　蒂：太可惜了！我剛剛還正在為妳們業績成長許多開心呢！

伊莉莎白：對呀！我也希望我們的訂單真的是訂 22 組哩！妳可以讓我們修改訂單嗎？

貝　　蒂：應該沒問題，因為也還沒安排出貨。

伊莉莎白：那我就放心了！真抱歉造成妳的不便！

貝　　蒂：沒關係的，下次妳發訂單來，我就會快快處理、快快出貨。我相信大單的量妳絕對還是能快快賣完的呢！

伊莉莎白：呵呵……以後在發訂單給妳之前，我會再檢查一次所有內容的！

電話英文短句─好說、說好、說得好！

- What a coincidence!　好巧！
- What happened?　怎麼了呢？
- That's too bad!　太可惜了！
- That should be alright.　應該沒問題。
- What a relief!　那我就放心了！

🌐 **國貿知識補給站** 當修改、取消訂單行不通時…

　　下訂單是令大家都開心的事，業務人員一段時間以來的辛勞終於告一段落，可讓自己業績數字更漂亮些，也能替自己掙得獎金報酬。廠商當然也開心，因為銷售金額又往上層疊增加，庫存壓力又小了些。不過，完美的事若是天天發生，哪裡是人生啊？！下了訂單之後，其實不時都會有那種突然發現有錯、足以讓人呆住楞住傻住憋住（我說的是憋氣）的關鍵五秒鐘，腦子裡不斷跑著「糟了、完了、怎麼辦？」的跑馬燈！那訂錯了到底該怎麼辦呢？跟廠商要求修改、取消，廠商若依了，就是你有求，廠商有應，緊急狀況馬上就可化解掉。那廠商若不依呢？我們現在就來看看廠商會有哪些個不依法、會有什麼樣的情境了！

接單生產，改不得，取消不得

　　在出貨前發現錯誤，通常會有好結果，因為廠商多半都可動動指頭，修一下訂購品項，改一下數量、出貨要求就可搞定。但是，對於接單生產的訂單，廠商依客戶需要而做，一旦下單，想要修改或是取消就不行了。廠商會這樣回你："It is a made-to-order item which cannot be canceled."（這是接單生產的品項，不得取消），或說 "This order is binding and it's not possible to cancel it."（這份訂單具約束性，無法取消。）當你看到這樣的回覆，就知道無力回天了，請容許自己黯然神傷一分鐘，接著就要再打起精神，積極思考如何銷售！

可以取消不出貨，但要收費

　　有的廠商雖然還沒出貨，但其實已自倉庫取貨、包裝，已做了出貨

的安排，對於這樣的情況，廠商確實可不出，但卻得跟你收取個費用。廠商話會是這麼説的：“we have to charge restocking fee if you cancel the order"（若你要取消訂單，我們得跟你收取重新入庫費。）。

出貨了⋯直接退運吧！

當你跟廠商修改或取消訂單時，發現廠商貨已出，那該如何是好？若是評估來貨實在難銷，仍決定退貨，就可跟海關申請「直接退運」，來貨不辦進口，直接退運出境，以省去相關稅負、省去進口後再辦退運出口所耗費的人力與成本。

出貨了，貨來了，賣賣看吧！

若是廠商已出貨，訂單要修要取消都來不及了，那就得好好計算一下把貨留下來的銷售勝算是如何。有的原廠不接受退貨，所以你也只能摸摸鼻子、硬著頭皮吃下失誤，吃下來貨。在這樣的情況下，可在付款條件上跟廠商協商一下，請求廠商不要馬上要求付款，等確實銷貨後，再開發票，再支付貨款。若是一段時間過去了，還是無法成功銷售出去，那就還是要辦理退貨了。我們來看一下廠商在碰到這樣的要求時，會有哪幾種可能的回覆：

➢ We are unable to re-sell any kits once they have left our site. For this reason we are unable to provide a refund for the extra kits. （貨一旦出了，我們就不能收回，重新再銷售，所以無法給我們這些多訂貨品的退款。）

➢ Please keep the product and pay after you sell it. （請留著貨，等銷售後再付款。）

➢ Please try and sell these extra kits. Whatever kits you have left by

the end of July, please return to us and we will then refund 50%. This means that we are both paying 50% for the error that has been made.（請試著銷售這些多訂的貨, 到七月底時, 若是貨還有剩，可退貨給我們，我們則會退給你 50%的貨款，換言之，對這次的訂貨失誤，我們雙方各自分擔 50%的損失。）⋯這絕對是有佛心來著的廠商，若是有幸遇到這樣的廠商，請務必記著這樣的善意，待日後任何人有任何出錯與失誤時，請也以同樣的誠心來共同解決問題，讓錯誤在達達的馬蹄聲中，一樣能有美麗的體悟！

《Dear Amy》時間

Dear Amy，

　　您好，我有一個問題，困擾我很久了，我想來問問您要怎樣才能在下單作業上更順一點？我負責跟國外廠商下單，廠商對於星期幾下單都有清楚的要求，但是每次我一發正式訂單給國外之後，業務就陸續來要求追單了，隔天早上追兩項，下午取消一項，再隔天還再追個三項，我就得一直跟國外改，國外也不止一次告訴我，請我們一次下足訂單，說我這樣一直追單他們會搞混…我是很想配合國外，可是同事又不配合我！請問我要怎樣才能讓業務下單後就不修不改？怎樣才能不會讓國外搞混？怎樣才能讓我自己做事快樂一點呢？

修改訂單修到沒力的 Vita

Dear Vita,

　　請不要沒力，我光是看妳的名字就覺得很有力、很帶勁兒呢！關於妳的困擾，我們先來想想：業務想不想一直修改訂單呢？如果一次可以做好的事，沒有人會想要分好幾次做，為自己添麻煩，所以，修改訂單這種要求，最終還是追溯到客戶身上，業務是可以免責的。那我們在體諒業務之餘，就沒辦法改變問題了嗎？有的，妳想想，若是國外廠商作風強硬，過了下單日之後就不准追單，客戶也就會在下單日一次下好訂單。我們若要在作業上達到這樣的境界，就要請業務教育客戶，從客戶的利益角度來說明，讓客戶清楚知道唯有一次下足訂單，才能確保完整出貨，不會亂掉漏掉。另一方面，若是真的碰到訂單得修個好幾次，怎樣才能讓國外不亂不漏呢？請在主旨清楚扼要寫出<u>要修改訂單、第幾次修改</u>這主題，在 e-mail 內文裡，則請寫出要<u>追訂幾項、追訂後整個訂單共有幾項</u>，讓國外對照一下數字就知道項次對否。最後要說的這個層面最重要了，要怎樣讓妳快樂？請想著追單代表公司業績又更好一點，是好事，我們說得讓國外不糊塗，是美事，再來，「Vita」代表了生命、生活，每個人在叫妳的時候，也都是灌注活力給妳，有這麼好的事，一定每天都要覺得快樂的啊！

單字片語說分明

• **amend** [ə`mend]

v. 修改 to make changes to a document, law, agreement etc., especially in order to improve it.

= revise, modify, correct, alter

例 Unfortunately we are unable to amend the price quotation that we sent last week. Sorry.

可惜我們無法修改上星期所發的報價單，抱歉。

amendment n 修訂；修正條款

make amends 賠償；補救

• **adjust** [ə`dʒʌst]

v. 調整 to change something slightly in order to make it better, more accurate, or more effective

例 I have signed the invoices as requested and, if you have any issues with customs, please advise so we may adjust accordingly.

我們已依您要求在發票上簽名了，若是海關還有其他要求，就請告訴我們，我們或許可配合調整。

其他字義：調節；適應

adjustable adj 可調整的

adjustment n 調整

make adjustments 做調整

• **cancel** [`kænsl]

v. 取消 to say that something that has been arranged will not now happen

例 Please find the attached credit memo, to cancel original Invoice IN040600, and revised Invoice IN040601 reflecting your correct discount code.

請見附件的貸項憑單，以取消原來的發票 IN040600，另也附上含正確折扣代碼的發票 IN040601。

cancellation ⓝ 取消；撤銷；廢止

· update [ʌp`det]

v. 更新 to add the most recent information to something such as a book, document, or list

例 After the holiday break, I will send an updated Quotation to you for the inquired instrument.

在假期過後，我會寄給您所詢儀器的更新報價單。

update ⓝ 更新；最新消息

· original [ə`rɪdʒənl]

adj. 原先的 existing at the beginning of a period or process, before any changes have been made

例 Attached is the invoice for this order. The original is to follow via the mail.

在此附上這份訂單的發票，正本將會以郵寄寄出。

其他字義：全新的；有獨創性的

original ⓝ 原件；原作

origin ⓝ 來源；出身

originality ⓝ 獨創性

originate ⓥ 發源；創始

· retain [rɪ`ten]

v. 保留 to keep someone or something

例 Here is the shipping notification for your order. Please retain for your records.

這是您訂單的出貨通知單，請留著做為記錄。

其他字義：記住

字尾- tain 集合！

-tain, ten = hold 握；持

單字	組合	字義
contain	con 共同＋tain 握（都握在一起）	包含；容納
maintain	main=man 手＋tain 握（握在手裡）	保持；維持
obtain	ob 加強意義＋tain 握（握緊）	取得
sustain	sus=sub 下＋tain 握（從下方握住）	支撐；維持

- **exception** [ɪkˋsɛpʃən]

n. 例外 someone or something that is different in some way from other people or things and so cannot be included in a general statement

例 We do not sell this product as single-boxes. However, if you feel that this is a potentially big customer and it is likely they will continue purchasing a 10-box size, I can make an exception.

這一項產品我們並不做單包銷售，不過，若是您覺得這是有潛力的大客戶，而且可能之後會購買整組十包的規格，那我們可破例處理。

exceptional adj 例外的

常見搭配詞、片語：

notable exception	rare exception	make an exception
明顯的例外	罕見的例外	破例
be no exception	with the exception of	without exception
不例外	除……之外	沒有例外地

in exceptional circumstances/ cases 在特殊情況下

- **postpone** [postˋpon]

v. 延期 to decide that something will not be done at the time when it was planned for, but at a later time

例 If there is not enough interest in the speech now, we may postpone to next year.

若是目前對這場演講並沒有太多的興趣，我們可以延到明年再辦。

字首 post- 集合！

post- = after 之後

單字	組合	字義
postmodern	post 之後＋modern 現代	後現代
postscript (p.s.)	post 之後＋script 筆跡	附筆；附記
postwar	post 之後＋war 戰爭	戰後

- suspend [sə`spɛnd]

v. 中止 to officially stop something for a short time

例 In the event that the project is suspended or delayed, the client will be billed for the work and expenses incurred.

如果這個專案中止或延後執行，那麼，已做的工作與產生的費用，皆須由客戶支付。

其他字義：責令…停職、停學；懸、吊、掛

suspension n 中止；停職、停賽；懸吊系統；懸浮液

suspension bridge 吊橋

suspense n 忐忑不安；懸疑

- circumstance [`sɚkəmˌstæns]

n. 情況 a fact or condition that affects a situation

例 It also looks like we won't be able to ship your another order due to circumstances that were discussed previously.

因為先前所說的那些情況，看來您的另一份訂單我們也沒辦法出貨了。

其他字義：境況；（正式）機運、命運

常見搭配詞、片語：

under/ in certain/ different/ exceptional circumstances
在某些／不同／特殊情況下

not under/ in any circumstances 無論如何都不會

due to unforeseen circumstances 由於事先無法預測到的情況

due to circumstances beyond our control 由於情況非我們所能控制

Part 4
銀貨兩訖篇

Unit 1 出貨
Unit 2 付款

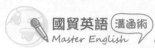

國貿英語 溝通術
Master English　Communication for International Trade

Unit 01

出貨

單字片語一家親

出！

出貨	shipment
運送	delivery
迅速寄送	dispatch
運送、運輸	transport
班機	flight
離開、出發	departure
目的地	destination
貨運	freight
出口	export
進口	import
出貨人	sender, shipper
收件人	recipient, consignee

貨要裝出了！

貨物	goods, cargo
包裝	package
易碎的	fragile
包裹	parcel
容器	container
裝貨箱；條板箱	crate
棧板	pallet
裝貨、上貨	loading
材積、尺寸	dimension

淨重	net weight
毛重	gross weight

與出貨方式有關…

空運	air freight
海運	sea freight, ocean freight
運輸業者	carrier
貨運公司	freight carrier
快遞	express, courier
優先型服務	priority service
經濟型服務	economy service
郵寄服務	postal service

文件資料要附！

空運提單	Air Waybill
主提單	Master Air Waybill
分提單	House Air Waybill
海運提單	Bill of Landing
商業發票	Commercial Invoice
包裝單	Packing List / Packing Slip
標籤	label, tag

最後通關

清關	customs clearance
取貨	collect, pick-up
關稅	tariff, customs duties

142

句型

句型 1 通知已出貨了

Your order has been shipped out via... and should arrive shortly.

例 Your order has been shipped out via your FedEx account today and should arrive shortly.

您的訂單已在今天以您聯邦快遞的帳號出貨，應該很快就可送達。

句型 2 出貨預告

Please be advised that ...

例 Please be advised that the available items will be sent out on Friday.

在此通知您，現貨品項將會在星期五出貨。

句型 3 有現貨可出

We currently have + N. + in stock and would be able to ship to you.

例 We currently have the kits in stock and would be able to ship to you by end of this week if you confirm the expiries ASAP.

這幾組目前有現貨，若是您能儘快確認效期一事，我們就可在週末前出貨給您。

句型 4 哪個方式合意？

Do you prefer to ...

例 Do you prefer to receive a partial shipment with the cell line only? Please advise.

您想要我們安排分批出貨，先出細胞株嗎？請告知。

句型 5 送上要求的資料

Per your request, please find + N. + attached.

例 Per your request, please find a copy of your invoice attached.

依您所要求，在此附上發票影本。

安排出貨 E-mail | Arranging Shipment

安排出貨 E-mail 範例 **1** · 通知出貨、提供出貨文件、提醒「稅」事

Dear Arthur,

Your order has been shipped out via FedEx AWB# 6107 9687 6992 and should arrive shortly. Per your request, please find the complete set of shipping documents attached. The original ones will be packed together with the shipment. All documents included on the outside of the package will have a signature as requested. Please let us know if you require anything further for customs clearance.

Please be advised that the recipient is responsible for all Duties & Taxes imposed by your local government/customs officials. In the event that Duties & Taxes are not collected by the shipping agency at the time of delivery, we will bill you the amount of the Duties & Taxes incurred.

Thank you for giving us the opportunity to assist you with your research. If you should have any questions regarding your shipment, please don't hesitate to contact us.

Sincerely,
Jessie Hsu
Summit Technologies, Inc.

單字 ShabuShabu 一小補小補！
☑ signature [`sɪgnətʃɚ] (n.)　簽名
☑ impose [ɪm`poz] (v.)　課徵
☑ incur [ɪn`kɝ] (v.)　帶來
☑ summit [`sʌmɪt] (v.)　高峰

安排出貨 E-mail 範例 1 中文

亞瑟，您好，

　　您的訂單已出貨，聯邦快遞提單號碼為 6107 9687 6992，應很快就可送達。依您所要求，在此附上完整的一份出貨文件，正本將會隨貨寄出，所有的文件上都會如您要求簽上名。如果您在清關上還有任何需要，再請您告知。

　　在此通知您，收件方要負擔所有政府／海關所課徵的關稅與稅負，若貨運代理公司在出貨時未收取關稅與稅負，我們會將衍生的關稅與稅負計入您帳上。

　　謝謝您給我們這個機會，能為您的研究提供協助，如果對於出貨還有任何問題，請儘管與我們連絡。

謹上

徐潔西
高峰科技公司

Check 好句

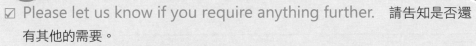

☑ Please let us know if you require anything further.　請告知是否還有其他的需要。

☑ Thank you for giving us the opportunity to assist you.　謝謝您給我們這個機會為您提供協助。

安排出貨 E-mail 範例 2 · 分批出貨、包裝安排

Dear Peter,

　　We have received your PO # 1031001. We currently have the ordered 1 kit of # 270 and 2 kits of # 271 in stock and would be able to ship to you this Friday. The kit of # 265, however, will not be in stock until August 16th. Do you prefer to receive a partial shipment now or shall we hold the order until August 16th? Please advise.

　　For your information, we'll ship it via FedEx international priority express, taking 3 days to arrive to you. It is shipped in 20 lb dry-ice package, dimension 13x12x11 inch box. We prefer to use your FedEx account for this shipment. If so, we'll only charge $50 for the box and dry-ice fee.

　　In the meantime, should you require any further assistance, please do not hesitate to contact me.

Best Regards,

Joyce Pan
Formosa Chemical

單字 ShabuShabu 一小補小補！
☑ partial [ˋpɑrʃəl] (adj.) 部分的
☑ dry-ice (n.) 乾冰
☑ meantime [ˋmin͵taɪm] (n.) 其間

安排出貨 E-mail 範例 2　中文

彼得，您好，

　　我們已收到您單號 1031001 的訂購單，我們目前有您所訂的一組 ＃270 與兩組 ＃271 的現貨，可在這個星期五出貨給您，不過，＃265 這一組要到八月十六日才會有貨，請問您要分批出貨嗎？還是要我們在八月十六日之前整批訂單先不出貨？請告知。

　　跟您說一聲，我們會安排走聯邦快遞國際優先型出貨，三天後會送達。這批是以 20 磅乾冰包裝來出貨，箱子的尺寸為 13x12x11 英吋，我們偏好以您的聯邦快遞帳號來辦理出貨，若是可這樣安排，那我們就只需要收取美金 50 元的箱子與乾冰費用。

　　在此同時，若您還有其他的地方需要協助，就請儘管直接與我連絡。

祝好

潘喬伊絲
福爾摩沙化學公司

Check 好句

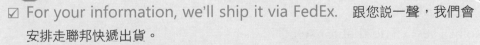

☑ For your information, we'll ship it via FedEx.　跟您說一聲，我們會安排走聯邦快遞出貨。

☑ In the meantime, should you require any further assistance, please do not hesitate to contact me.　在此同時，若您還有其他的地方需要協助，就請儘管直接與我連絡。

電話對話

電話詢價與報價 範例

人物介紹

Kelly

凱莉，原廠客服人員，辦事效率佳，對於交派的工作，總是能辦得又快又好！

Rita

莉塔，業務秘書，行事謹慎，能將事情安排得滴水不漏！

Kelly: Hello, Kelly Jackson speaking. How may I help you?

Rita: Hi, Kelly. This is Rita from Dynasty Technologies in Taiwan. I'm calling to follow up on an e-mail that I sent you yesterday.

Kelly: Yeah, I'm currently working on your request about bottling. I just talked to our Production Manager. Normally we do not use amber glass bottles for your ordered volume, but we can manage to arrange for you.

Rita: Thanks indeed! We want to assure that the product quality will not be affected during transport.

Kelly: Sure! The quality is always our first priority.

Rita: There's just one more thing I'd like to remind you. Please pack together the Invoice and product's data sheet with the shipment.

Kelly: Not a problem! I'll do that.

Rita: Great! We need these documents so as to clear the goods from our customs.

Kelly: I see. I'll do that for sure!

Rita: Thanks! I know I can always count on you!

Kelly: Hahhah… Trust me! I'll never let you down!

PART
1

PART
2

PART
3

PART
4

PART
5

PART
6

PART
7

電話修改訂單 範例 中文

凱莉：您好，我是凱莉‧傑克森，請問有什麼我能幫忙的地方嗎？

莉塔：嗨，凱莉，我是臺灣朝代科技的莉塔，我打來是要跟您問問我昨天 e-mail 所說的那件事。

凱莉：我知道，我正在處理妳提的裝瓶要求這事呢！我剛才有跟我們的生產部經理談過，通常，對於妳所訂的這個量，我們並沒有用茶色玻璃瓶來裝貨，不過，我們可以想辦法來配合安排。

莉塔：真的很謝謝妳！我們想要確保產品品質不會在運送途中受到影響。

凱莉：當然！品質一向是我們的首要堅持。

莉塔：還有一件事我要提醒妳一下，請出貨時同時附上發票及產品說明書喔！

凱莉：沒問題！我會照辦。

莉塔：太棒了！我們清關時就得要用到這些文件。

凱莉：了解，我一定會照辦！

莉塔：謝啦！我就知道靠妳準沒錯！

凱莉：哈哈！相信我，我決不會讓妳失望的！

電話英文短句－好說、說好、說得好！

- Thanks indeed! 真是太謝謝您了！
- Sure! 當然！
- There's just one more thing I'd like to remind you.
  還有一件事我要提醒你一下。
- Not a problem! I'll do that. 沒問題！我會照辦。
- I know I can always count on you! 我就知道靠你準沒錯！
- Trust me! 相信我！

國貿知識補給站 出貨方式面面觀

　　下了訂單之後，殷切期盼的就是來貨的那一刻！國際貿易路途遙遠，飛越天際又橫渡海洋，從出貨到送達目的地，中間可有個好幾個關卡，每一關可也都需要紮實、仔細的功夫，才不會終於盼得原廠出貨，卻發生這兒卡那兒卡的不順狀況，白白延誤了來貨的預定時程。

　　要讓貨物橫越國際線，若以承接貨物、辦理通關的中介角色來分，則可分出一般海空運、快遞公司的海空運，以及郵局的快捷與一般航空、水陸、陸空聯運。就不同空運的速度來看，快遞快過一般空運，一般空運與郵局快捷差不多，郵局快捷又快過其一般航空，而在運費上，要快當然索價就高。要如何選擇，就要看你的成本考量與時間要求的條件了。

　　運送方式不同，連有關聯的各個角色的稱呼也就不一樣了呢！請你瞧瞧下表各路人馬的稱號：

出貨方式	賣方	運送方	買方
一般海空運	Shipper (託運人；寄件人)	Carrier (運送人) Forwarder (貨代)	Consignee (受託人；收件人)
快遞公司	Sender (寄件人)	Courier (快遞公司)	Recipient (收件人)
郵局	From (從；寄件人)	Post office (郵局)	To (到；收件人)

　　在出貨的安排上，首先客戶須將收貨的基本資訊提供給原廠，包括出貨地址、電話與連絡人這些基本的身家資料。接著，原廠就要知道你想要他們怎麼出貨，走快遞、一般型、還是郵局呢？

　　若是走快遞，收件人則須讓原廠知道你有無什麼偏好的快遞公司、你有無申請快遞公司的帳號？如有，則出貨的運費就會是以快遞公司跟你簽約的條件來計算。寄件人要做的工作為包裝、準備託運文件，

也就是填寫空運提單（Air Waybill）與商業發票（Commercial Invoice），隨貨附上，接著連絡快遞公司到寄件人處收件（pickup），完成交寄作業。貨一送出，就是辦理通關、清關（clearance）作業了，再過了這一關，收件人就可以等著貨物送上門囉（delivery）。

　　若客戶要求原廠安排一般空運，則原廠會這麼要求："If you choose a freight forwarder option, please note that you will be responsible for coordinating all paperwork and scheduling the pickup."（若是你選擇透過貨物承攬業務代理公司（簡稱貨代）辦理出貨，請注意您就必須負責安排所有的書面作業與取貨事宜。）客戶在決定選擇哪一家公司為貨代時，通常會要求原廠提供一個訊息："Please let us know the gross weight and dimensions of the shipment."（請告知此批貨的毛重與材積），有了這兩個數據，才能「費」比三家，找出最划算、最佳的貨代。運費的計算基礎通常是看毛重與總材積重量，以較重者為準。決定好貨代之後，就可連絡貨代前往原廠取貨了。原廠所要準備的出貨文件，則包括了提單（Air Waybill）、發票（Invoice）與包裝單（Packing List、Packing Slip）。

　　若要透過郵局來出貨，以我國的中華郵政為例，其提供了一般國際包裹，以及國際快捷的服務，國際快捷亦稱 EMS（International Express Mail Service），寄件人須填寫五聯單與商業發票，至郵局窗口辦理，出貨後有 EMS 號碼可追蹤交寄的貨物。不過，郵局寄送服務限制較多，有禁寄物品的規定，還有尺寸限制與重量限制，在重量部分，依寄送國家不同，限重 20 公斤或 30 公斤。有時還會看到美國廠商通知說他們透過 USPS 出貨，乍看會以為是 UPS 美商優比速快遞

（United Parcel Service of America），其實 USPS（United States Postal Service）指的也就是美國郵局。

運送方式有好幾種，可選擇的運輸公司更是有好多家，及早找到固定的配合業者，除了可談到好費率，配合上也能更有默契，如此一來，就能人順、事順、貨物一路順呢！

出貨文件 — 空運提單 Air Waybill 範例

297 LAX 8252 1000

Shipper's Name and Address CHEMPORT CORP. 1000 XXXX AVE CHINO, CA 91710	Shipper's Account Number	**Not Negotiable** **Air Waybill** issued by Ⴘ RIGHT FREIGHT INC.
		Copies 1, 2 and 3 of this Air Waybill are originals and have the same validity
Consignee's Name and Address BIOCHEM SCIENCE, INC. 37F, NO. 337, RUIGUANG RD., NEIHU DIST., TAIPEI, 114, TAIWAN	Consignee's Account Number	It is agreed that the goods declared herein are accepted in apparent good order and condition (except as noted) for carriage SUBJECT TO THE CONDITIONS OF CONTRACT ON THE REVERSE HEREOF. ALL GOODS MAY BE CARRIED BY ANY OTHER MEANS INCLUDING ROAD OR ANY OTHER CARRIER UNLESS SPECIFIC CONTRARY INSTRUCTIONS ARE GIVEN HEREON BY THE SHIPPER, AND SHIPPER AGREES THAT THE SHIPMENT MAY BE CARRIED VIA INTERMEDIATE STOPPING PLACES WHICH THE CARRIER DEEMS APPROPRIATE. THE SHIPPER'S ATTENTION IS DRAWN TO THE NOTICE CONCERNING CARRIER'S LIMITATION OF LIABILITY. Shipper may increase such limitation of liability by declaring a higher value for carriage and paying a supplemental charge if required
Issuing Carrier's Agent Name and City Ⴘ RIGHT FREIGHT INC. 2000, XXXX BLVD., INGLEWOOD, CA 90301		Accounting Information HOPING FREIGHT LTD. 5F, NO. 5, YONG JI RD., XIN YI DISTRICT 110 TAIPEI TAIWAN TEL: (02)2700-1000 FAX: (02)2700-1001 ATTN.: DAVID//TERESA/WENNY
Agent's IATA Code 01-1-9000/0010	Account No.	

Airport of Departure (Addr. of First Carrier) and Requested Routing LOS ANGELES AIRPORT						Reference Number			Optional Shipping Information		

To TPE	By First Carrier CI5100/10	Routing and Destination	to	by	to	by	Currency US$	CHGS	WT/VAL		Other		Declared Value for Carriage	Declared Value for Customs
									PPD	COLL	PPD	COLL		

Airport of Destination CKS AIRPORT	Requested Flight/Date 12/12/2014	Amount of Insurance	INSURANCE - If carrier offers insurance, and such insurance is requested in accordance with the conditions thereof, indicate amount to be insured in figures in box marked "Amount of Insurance".

Handling Information		
		SCI

No. of Pieces RCP	Gross Weight	kg lb.	Rate Class		Chargeable Weight	Rate/Charge	Total	Nature and Quantity of Goods (incl. Dimensions or Volume)
				Commodity Item No				
2 CTNS	50L 25K			50L 25K				LAB RESEARCH REAGENTS (RESEARCH PURPOSE ONLY) "FREIGHT PREPAID" THIS SHIPMENT IS SUBJECT TO INSPECTION DRY ICE NEEDS SPECIAL HANDING BOX 1- PLS KEEP FROZEN

Prepaid	Weight Charge	Collect	Other Charges
			PREPAID
Valuation Charge			
Tax			I hereby certify that the particulars on the face hereof are correct and that insofar as any part of the consignment contains dangerous goods. **I hereby certify that the contents of this consignment are fully and accurately described above by proper shipping name and are classified, packaged, marked and labeled, and in proper condition for carriage by air according to applicable national governmental regulations.**
Total Other Charges Due Agent			
			CHEMPORT CORP.
Total Other Charges Due Carrier			Signature of Shipper or his Agent
			AS AGENT FOR
Total Prepaid	Total Collect		12/12/2014 LOS ANGELES BRIGHT FRIGHT INC.
Currency Conversion Rate	CC Charges in Dest Currency		Executed on (date) at (place) Signature of Issuing Carrier or its Agent
For Carrier's Use only at Destination	Charges at Destination	Total Collect Charges	**BFI-12200**

　　空運提單真格是份大單！許多的欄位與資訊就這樣塞進一張單子裡，它「排」「擠」得辛苦，我們可也得好好地、細細地來瞧一瞧，到底它英文説啥，中文是何意，也讓只看少數幾個特定欄位的大多數人，終於有個機會補足那一直以來的空白，一字一句地走一趟提單大地圖裡的每個角落。

一、左上區塊

297 LAX 8252 1000

Shipper's Name and Address 出貨人名稱與地址	Shipper's Account Number 出貨人帳號
CHEMPORT CORP. 肯博公司 1000 XXXX AVE　1000 XXXX 街 CHINO, CA 91710　奇諾，加州 91710	
Consignee's Name and Address 收貨人名稱與地址 BIOCHEM SCIENCE, INC. 37F, NO. 337, RUIGUANG RD., NEIHU DIST., TAIPEI, 114, TAIWAN	Consignee's Account Number 收貨人帳號 拜爾肯科學公司 臺灣臺北市內湖區瑞光路 337 號 37 樓
Issuing Carrier's Agent Name and City　開立提單之承運人代理名稱與地址 B RIGHT FREIGHT INC.　　明亮貨運公司 2000, XXXX BLVD.,　　美國加州英格塢 XXXX 大樓 2000 號 INGLEWOOD, CA 90301	
Agent's IATA Code 代理人之國際航協代號 01-1-9000/0010	Account No. 帳號

Airport of Departure (Addr. of First Carrier) and Requested Routing 出發機場（啟運運送公司之地址）與要求路徑 LOS ANGELES AIRPORT 洛杉磯機場							
To TPE 至 臺北	By First Carrier CI5100/10 透過啟運航空公司 CI5100/10	Routing and Destination 路線與目的地	to 前往	By 透過	To 前往	By 透過	
Airport of Destination 目的地機場 CKS AIRPORT　桃園機場		Requested Flight/Date 要求之航班／日期					
		12/12/2014					

二、右上區塊

Not Negotiable 不具流通性 **Air Waybill** 空運提單 issued by 提單開立是由 ＢRIGHT FREIGHT INC. 明亮貨運公司
Copies 1, 2 and 3 of this Air Waybill are originals and have the same validity 此空運提單第一、二、三份為正本聯，具同等效力。
It is agreed that the goods declared herein are accepted in apparent good order and condition (except as noted) for carriage SUBJECT TO THE CONDITIONS OF CONTRACT ON THE REVERSE HEREOF. ALL GOODS MAY BE CARRIED BY ANY OTHER MEANS INCLUDING ROAD OR ANY OTHER CARRIER UNLESS SPECIFIC CONTRARY INSTRUCTIONS ARE GIVEN HEREON BY THE SHIPPER, AND SHIPPER AGREES THAT THE SHIPMENT MAY BE CARRIED VIA INTERMEDIATE STOPPING PLACES WHICH THE CARRIER DEEMS APPROPRIATE. THE SHIPPER'S ATTENTION IS DRAWN TO THE NOTICE CONCERNING CARRIER'S LIMITATION OF LIABILITY. Shipper may increase such limitation of liability by declaring a higher value for carriage and paying a supplemental charge if required 茲同意在此所申報的貨品，依照背面運送契約條款所規定，送交的貨物皆具良好狀況。所有的貨物皆可透過包括公路運輸等方式運送，或者，皆可由任何其他運送人運輸，除非托運人有不同的指示。此外，托運人同意貨物可運送至運送人認為合適的中途停留點。另請托運人注意，運送人的責任有限，如有需要，托運人可藉由對運送人申報較高的價值並支付附加費用，以擴充此責任限度。
Accounting Information 帳務資訊 HOPING FREIGHT LTD. 希望貨運有限公司 9F, NO. 5, YONG JI RD., XIN YI DISTRICT 110 TAIPEI TAIWAN 臺灣臺北市信義區永吉路 5 號 9 樓 TEL: (02)2700-1000　FAX: (02)2700-1001　　電話：(02)2700-1000 傳真：FAX: (02)2700-1001 ATTN.: DAVID//TERESA//WENNY 收件人：大衛／德瑞莎／溫妮

Reference Number 參考號碼				Optional Shipping Information 其他出貨資訊			
Currency 幣別 US$ 美金	CHGS 付款方式	WT/VAL 航空運費／聲明價值附加費		Other 其他		Declared Value for Carriage 供運用所聲明之價值	Declared Value for Customs 供海關用所聲明之價值
		PPD	COLL	PPD	COLL		
Amount of Insurance 保險金額		INSURANCE - If carrier offers insurance, and such insurance is requested in accordance with the conditions thereof, indicate amount to be insured in figures in box marked "Amount of Insurance". 保險一若運送人有提供保險，而此保險為依其條件所要求，則請在箱子上註明保險金額為何，標註「保險金額」。					
Handling Information 處理資訊		SCI 海關資訊					

三、下半區塊

No. of Pieces RCP 件數 運價點	Gross Weight 毛重	kg lb. 公斤 鎊	Rate Class 運價種類 / Commodity Item No 貨品型號	Chargeable Weight 計費重量	Rate/Charge 費率	Total 總計	Nature and Quantity of Goods (incl. Dimensions or Volume) 貨物之品名與數量（包含材積或體積）
2 CTNS 2 個紙板箱	50L 25K 50 公升 25 公斤		50L 25K 50 公升 25 公斤				LAB RESEARCH REAGENTS (RESEARCH PURPOSE ONLY) 實驗室研究用試劑（僅供研究用）

Prepaid 預付	Weight Charge 計費重量	Collect 到付	Other Charges PREPAID 其他費用 預付
	Valuation Charge 估價費用		
	Tax 稅		I hereby certify that the particulars on the face hereof are correct and that insofar as any part of the consignment contains dangerous goods. **I hereby certify that the contents of this consignment are fully and accurately described above by proper shipping name and are classified, packaged, marked and labeled, and in proper condition for carriage by air according to applicable national governmental regulations.**
	Total Other Charges Due Agent 代理人應付之其他費用總計		
			茲在此證明提單正面所記述內容正確，並在委託貨物含危險物品之規定範圍內。茲在此證明此委託貨物的內容已完整、正確描述如上，記有其合適的出貨名稱、分類、包裝、標註與貼標，根據適用的國家政府規定，貨物亦適於空運運輸。
	Total Other Charges Due Carrier 運送人應付之其他費用總計		
			CHEMPORT CORP. 肯博公司
			Signature of Shipper or his Agent 托運人或其代理人簽名
Total Prepaid 預付總計	Total Collect 到付總計		AS AGENT FOR 代理人 LOS ANGELES BRIGHT FRIGHT INC. 洛杉磯 明亮貨運公司
Currency Conversion Rate 幣別 匯率	CC Charges in Dest Currency 目的站到付費用		
			Executed on (date) 執行日（日期）　at (place) 執行地（地點） Signature of Issuing Carrier or its Agent 製單運送人或其代理人之簽名
For Carrier's Use only at Destination 僅供運送人於目的站時使用	Charges at Destination 目的站費用	Total Collect Charges 到付費用總計	BFI-12200

出貨文件 一 發票 Invoice 範例

Invoice

Invoice Number: SIS10501
Invoice Date: 01/02/15
Page: 1

Bill to:	Ship to:
DRC International	DRC International
33F, No. 333, Ruiguang Rd., Neihu Dist., Taipei City 114, Taiwan	33F, No. 333, Ruiguang Rd., Neihu Dist., Taipei City 114, Taiwan

Sales Order Nbr:	SO1220	Ship Date:	01/02/15
Customer P/O Nbr:	10301014	Method:	FOB Dedham
Order Date:	01/02/15	Ship Via:	FEDEX
Salesman:	K RAMACHAN	Currency:	
Terms:	Net 30	AWB:	

Item No. Description	Ship Qty.	Size	Unit Price	Discount Pct	Total Price (USD)
AIG-00399 MAX Enzyme	5	100 units	500	10	2,250
Freight	1		50		50

Thank you for your order! All past due invoices are subject to a 1.5% monthly late charge. All prices include any applicable discounts and subvention obligations.	**Subtotal:** **2,300 USD** **Discount:** **Total:** **Tax:**

出貨文件 一發票 Invoice 範例　中文

發票

發票號碼： SIS10501
發票日期： 01/02/15
頁數： 1

付款人：	收件人：
迪爾希國際公司	迪爾希國際公司
臺灣臺北市內湖區瑞光路 333 號 33 樓 (114)	臺灣臺北市內湖區瑞光路333號33樓 (114)

銷售單號碼：	SO1220	出貨日期：	01/02/15
客戶訂貨單號碼：	10301014	出貨方式：	FOB Dedham
訂單日期：	01/02/15	出貨經由：	聯邦快遞
銷售人員：	K RAMACHAN	貨幣：	
付款條件：		提單：	

型號 品名	出貨數量	規格	單價	折扣率	總金額 (美金)
AIG-00399 麥克司酵素	5	100 單位	500	10	2,250
運費	1		50		50

謝謝您的訂單！ 所有逾期發票皆須收取1.5%之每月遲付費用。 所有價格皆已含任何可適用之折扣及補貼金。	小計： 2,300 USD 折扣： 總計： 稅：

國貿英語 溝通術
Master English Communication for International Trade

《Dear Amy》時間

Dear Amy，

　　我在一家進口代理公司上班，我們代理的品牌從日本、大陸、歐洲到美加都有，所以詢問出貨、追貨、處理通關狀況的事經常發生，但是我覺得我的工作情緒被這些狀況攪得有點糟，害得我原本已不怎麼 OK 的男朋友，在我沒啥心情、沒辦法好聲好氣之下，可能也快分手了！我工作上的情緒來自於我覺得我不斷地在處理同樣的問題，國外出貨了但沒來 Invoice，發 e-mail 去要，等個一天，來了卻發現金額與實際不符，有低報問題，又要再發 e-mail，又是一天過去了，業務急，我也煩…為什麼國外就不肯好好將所有正確的資料一次送來，讓我們可以好好報關呢？請問這樣的問題有沒有什麼方法可以終結呢？

工作愛情兩不順的 Monika

- -

Dear Monika,

　　若妳真覺得不順到極點，那就表示快要翻轉回順啦！出貨這事兒確實各種狀況都有，像是原廠貨出錯了、快遞公司將貨寄丟了、原廠說有出但我們就是沒收到等等。妳信中說到了出貨文件的問題，要是原廠做好一點，不要出錯就好了啊！是啊！但是別人是我們控制不來的，所以最好是反求諸己，想想我們自己要怎樣要求，才能讓原廠順著我們的意走，才好讓貨順順地到手呢？沒有別的訣竅了，就是請一次把話說清楚！若是妳要原廠發來 Invoice，以能趕快辦理通關，除了說明什麼時間之前一定要收到文件，接著就請好好說說妳要的 Invoice 內容，像是要列出品項細目、金額須為實際金額、不能有金額為「0」的品項。其實，要文件的人不喜歡一要再要，給文件的人也不喜歡一修再修啊！我們把話一次說足，事情就可能一次就辦妥，妳的心情就會好，看事看人也就都能更順、更客觀，就算工作老是有狀況出現，但請記著，關關難過最後還不是關關過？！所以，就請平心處事，待練成這道心法，不管是海關還是情關，必能關關好過輕鬆過！

PART
1
PART
2
PART
3
PART
4
PART
5
PART
6
PART
7

單字片語說分明

- shipment [`ʃɪpmənt]

  n. 出貨；運輸 the process of taking goods from one place to another

  例 Please be advised we will make shipment of your order the following Monday.

  在此通知您，我們將在下個星期一為您的訂單安排出貨。

  其他字義：運送的貨物

- shipping n. 運輸；運費

  與出貨有關的 shipping 家族：shipping ＋ 名詞

shipping documents　出貨文件	shipping agent　貨運代理商
shipping order　出貨單	shipping notice　出貨通知單
shipping container　貨櫃；運送箱	shipping schedule　出貨計畫

- delivery [dɪ`lɪvərɪ]

  n. 送貨 the process of bringing goods or letters to a place

  例 For up-to-date information about the delivery time for your order, please visit your account on our website.

  請至我們網站上登入您的帳戶，就可知悉關於您訂單出貨安排的最新消息。

  其他字義：提供服務；分娩；（電腦）傳輸

  deliver [dɪ`lɪvɚ] Ⓥ

- dispatch [dɪ`spætʃ]

  v. 派送；發送 to send someone or something somewhere

  例 Please let me know as soon as possible so I can get the paperwork to our warehouse to dispatch your order.

  請儘快告知，這樣我才能將書面作業的文件送到倉庫去，以發送您訂單的貨。

其他字義：（迅速地）處理掉

- transport [`træns͵pɔrt]

    n. 運送；運輸 the process of moving people or thing from one place to another

    = transportation

    例 Our company can transport goods more efficiently via a combination of rail, road, and sea.

    我們公司運用鐵路、公路、海路方式的搭配來運送貨物，效率更佳。

    其他字義：交通工具；運輸業

- container [kən`tenɚ]

    n. 貨櫃 a very large metal or wooden box designed to be loaded easily onto ships and trucks

    例 We put our products in containers and ship them around the world.

    我們的產品以貨櫃運輸，送到世界各地去。

    其他字義：容器

    contain [kən`ten] Ⅴ 裝有；容納；包含

    常見搭配詞

    Full-Container-Loads (FCL) shipping 整櫃運送

    Less-Than-Container Loads (LCL) Shipping 併櫃運送

    airtight container 氣密式容器

    sealed container 密封容器

    waterproof container 防水容器

· recipient [rɪˋsɪpɪənt]

n. 收件人 someone who receives something

例 If you are not the intended recipient, you hereby are notified that any dissemination, distribution, or copying of this communication is strictly prohibited.

如果您不是預定的收件人，在此通知您，任何傳播、散佈、抄寫此溝通訊息的行為，皆是嚴格禁止的。

receive [rɪˋsiv] ⓥ 收到；接受、接待

receiver [rɪˋsivɚ] ⓝ 收件人；電話聽筒；接收器

reception [rɪˋsɛpʃən] ⓝ 接受；接待；歡迎會

receptive [rɪˋsɛptɪv] ⓐⓓⓙ 樂於接受的

receptionist [rɪˋsɛpʃənɪst] ⓝ 接待員；櫃檯人員

· dimension [dɪˋmɛnʃən]

n. 材積；尺寸大小 the size of something

例 Each box would be 20 pounds with the dimensions of 18 x 16 x 11 (inches).

每一個箱子最多 20 磅，尺寸大小為 18 x 16 x 11（英吋）。

其他字義：維（長度、寬度或高度）

→ 3D Space：3-Dimensional Space（3 維空間）

· freight [fret]

n. 貨運 the transport system that carries goods

例 Please sign and return the Order Confirmation, so we'll get this order ready to ship by air freight.

請簽回訂貨確認單，這樣我們就可安排空運出貨。

其他字義：（運輸的）貨物；運費

常見搭配詞

air freight 空運

sea/ ocean freight 海運
freight carrier/ company 貨運公司
freight prepaid 運費預付
freight collect 運費到付

- economy [ɪˋkɑnəmɪ]

n. 節約；經濟 the careful use of money, products, or times so that very little is wasted

例 Please let me know if you would like me to solicit a FedEx International Economy Air or Sea Freight quote for you or if you plan to make your own arrangements.

請告知您要我向聯邦快遞要求報來國際經濟型空運或海運的運費嗎？或是您打算自行安排呢？

其他字義：經濟體制

economic [͵ikəˋnɑmɪk] adj 經濟的

economical [͵ikəˋnɑmɪk!] adj 經濟實惠的；節約的

economist [iˋkɑnəmɪst] n 經濟學家

常見搭配詞

economies of scales 經濟規模

make economies 節約

false economy 看似省錢但實際上更花錢的事物

boost the economy 促進經濟

stimulate the economy 刺激經濟

revive the economy 振興經濟

· collect [kə`lɛkt]

n. 取物;接人 to go and get a thing or person

例 Shipping charges will be added to the invoice, unless a FedEx account # or alternative collect shipping # is provided.

運費會加計到發票金額裡,除非您有提供聯邦快遞的帳號,或是其他的運費到付帳號。

其他字義:收集;收錢;贏得;積聚

collective [kə`lɛktɪv] adj 共同的;集體的

collectable [kə`lɛktəbḷ] adj 有收藏價值的;可收集的

collection [kə`lɛkʃən] n 收集;收藏品;一批;募集的錢

常見搭配詞

freight collect 運費到付

collect call 對方付費電話

collect your thoughts 定下心;專注

collective bargaining (勞資雙方就工資、工作條件進行的)集體談判

collective unconscious 集體潛意識

collector's item 值得收藏的東西;珍品

Unit 02

付款

 單字片語一家親

付款條件與方式

信用狀　letter of credit

付款交單　document against payment (D/P)

承兌交單　acceptance against payment (D/A)

電匯　wire, wire transfer telegraph transfer (T/T)

銀行轉帳　bank transfer

匯款　remittance

電子存款　electronic deposit

信用卡　credit card

銀行支票　bank check

什麼時候付？付多少？

預先付款　advance payment payment in advance prepayment up-front payment

到貨後付款　payment upon receipt of shipment

淨 30 天付款　net 30 days

月結付款　monthly payment

一次付款　full payment

部分付款　partial payment

分期付款　installment payment

頭期款　down payment

匯款事

申請人　applicant

受益人　beneficiary

銀行分行　branch

銀行帳戶名　bank account no.

銀行國際代碼　SWIFT code

匯款通知　remittance advice

手續費　bank fee wire transfer fee

信用卡事

信用卡持有人　card holder

卡別　card type

信用卡效期　good through credit card expiration date

三碼驗證碼 3 digit security code

授權簽名　authorized signature

信用卡授權表格 credit card authorization form

貨幣事

幣別　currency

匯率　exchange rate

匯兌　currency exchange

幣值　currency value

 **句型**

句型 1 順道一提

Taking this opportunity, we'd like to...

例 <u>Taking this opportunity, we'd like to</u> thank you for your great support.

藉此機會，我們要謝謝您的大力支持。

句型 2 提出建議

We would suggest changing to sth.

例 <u>We would suggest changing to</u> a different strategy of <u>marketing.</u>

我們建議改用另一個不同的行銷策略。

句型 3 懇請考慮

Please consider our proposal...

例 <u>Please consider our proposal</u> carefully and if you would like to comment, please complete the enclosed questionnaire.

請仔細考慮我們的提案，若有何意見，請填寫附件的問卷。

句型 4 關於提議嘛……

With regard to your suggestion of sth.

例 <u>With regard to your suggestion of</u> changing the ingredients of this product, would you please tell us why you think it would be better?

關於您所說更改成分的提議，可以請您告訴我們為何您會認為這樣比較好呢？

句型 5 佛心來著

However, in good faith, we'll honor sth.

例 <u>However, in good faith, we'll honor</u> the original quoted price.

不過，基於誠意，我們會依照原來的報價報給你。

PART 1
PART 2
PART 3
PART 4
PART 5
PART 6
PART 7

國貿英語 (溝通術)
Master English Communication for International Trade

付款條件討論 E-mail | Discussing Payment Terms

付款條件討論 E-mail 範例 **1** · 提議由預匯改為月結 30 天

Dear Tom,

We've made a remittance for the order that we placed on Monday. Attached please find the remittance advice for your reference. Taking this opportunity, we'd like to discuss the payment terms with you.

We've pleasantly cooperated with you for one year. As the number of our orders is increasing, the terms of advance payment for each individual order are not quite adequate for our present state considering the administrative time needed and bank fee incurred. So we would suggest changing to monthly payment. We'll make wire transfer in the mid of every month for the Invoices dated in the last month, e.g. we'll pay in mid April for all Invoices dated in March.

Please consider our proposal and confirm with us if it is acceptable to you. If you have any questions or concerns, please feel free to contact us. Thanks.

Best regards,

May Chang
Crystal Technology

單字 ShabuShabu 一小補小補！
☑ Individual [ˌɪndəˈvɪdʒʊəl] (adj.) 個別的
☑ adequate [ˈædəkwɪt] (adj.) 適當的
☑ present [ˈprɛznt] (adj.) 目前的
☑ administrative [ədˈmɪnəˌstretɪv] (adj.) 行政的

付款條件討論 E-mail 範例 1 中文

湯姆，您好，

　　關於我們星期一所下的訂單，我們已安排匯款，在此送上匯款通知書如附，供您參考。藉此機會，我們想要跟您討論一下付款的條件。

　　我們已跟您配合了一年，合作愉快，我們訂單的量也已增加了，若將行政處理所需的時間與產生的銀行手續費考慮進來，每次訂單都要預先付款的條件已不是那麼適合我們目前的情況了。因此，我們建議將付款條件改為月結，我們會在每月月中支付上一個月所開立的發票，例，我們會在四月中支付所有三月份的發票。

　　請考慮一下我們的提議，並與我們確認您是否可接受。若是您有任何的問題或考量，請儘管與我們連絡。謝謝。

祝好

張瑪莉
水晶科技

Check 好句

☑ Attached please find the remittance advice for your reference.　在此送上匯款通知書如附，供您參考。

☑ We've pleasantly cooperated with you for one year.　我們已跟您配合了一年，合作愉快。

回覆付款條件討論回覆 **E-mail** 範例 **2** · 雖無先例，但願意配合改為月結

Dear Mary,

Thanks for the remittance advise. Our bank notified us that we have now received the money into our bank account. Thanks.

With regard to your suggestion of changing the payment terms to monthly payment, our company policy is to have the prepayment confirmed prior to shipping. What we have done with other customers in the past is accepting a deposit ($5,000, for example) which we hold, and then we ship and make deduction on monthly payment terms. However, in good faith, we'll honor your proposed payment term. Mid-month for payment is fine, too.

However, on all wire transfers, please be reminded to indicate invoice numbers. It is too difficult to figure out which ones you are paying without the invoice number reference. Please do so accordingly. Thanks.

Best regards,

Tom Eden
Sunshine Enterprise

單字 ShabuShabu 一小補小補！
☑ notify [ˋnotəˌfaɪ] (v.)　通知
☑ deposit [dɪˋpɑzɪt] (n.)　存款
☑ indicate [ˋɪndəˌket] (v.)　指出
☑ accordingly [əˋkɔrdɪŋlɪ] (adv.)　照著

PART
1

PART
2

PART
3

PART
4

PART
5

PART
6

PART
7

回覆付款條件討論回覆 E-mail 範例 2　中文

瑪麗，您好，

　　謝謝您發來匯款通知，我們的銀行已通知我們此筆錢已入帳，謝謝。

　　關於您建議更改為月結的付款方式，我們的公司政策的規定是出貨前完成預付。我們過去有跟其他家公司配合過一種方式，就是接受他們的一筆存款（例如美金 5,000 元），留在我們帳上，然後我們就可依月結的方式來出貨與扣款。然而，基於誠意，我們可依照您所提的付款條件來做，也接受您每月月中付款。

　　不過，在此提醒您一下，所有的匯款都請列出發票號碼，因為，若沒有發票號碼供參考，很難理出究竟您支付的是哪些發票，還請您依此辦理，謝謝。

祝好

湯姆・伊登
日光企業

Check 好句

☑ Please be reminded to indicate invoice numbers.　在此提醒一下要列出發票號碼。

☑ Please do so accordingly.　還請您依此辦理。

電話對話

電話問貨、問匯款事 範例

人物介紹

Patricia

派翠西亞，原廠業務助理，記性好，資料查得快，應對不慌不亂。

Lucy

露西，業務秘書，有些急性子，但還懂得理與禮。

Lucy: Hello, Patricia. This is Lucy from Gamma Science.

Patricia: Hi, Lucy. What can I do for you?

Lucy: We have an active order with you. We've paid for it by wire transfer, but still haven't heard any further news about its shipment.

Patricia: Yeah, I remember you gave me your order last Monday. Let me check its status… Hmm…We haven't shipped indeed, but it's because we haven't receive your payment.

Lucy: Really? We wired last week!

Patricia: I see. Please let me explain the situation to you.

Lucy: I'm listening.

Patricia: It normally takes up to 2 weeks from the time a wire is made to actually get deposited into our account.

Lucy: Really? But we do need the goods urgently! Could you go ahead and ship for us today?

Patricia: I'm sorry. Our company policy is to ship only after receiving payment. I'll track your wire transfer with our bank and ship as soon as we get the money.

Lucy: OK. Please do keep an eye on it for us.

Patricia: I'll call you back after I talk to our bank.

Lucy: Thanks!

PART 1
PART 2
PART 3
PART 4
PART 5
PART 6
PART 7

電話問貨、問匯款事 範例 中文

露　　西：哈囉，我是珈瑪科技的露西。

派翠西亞：嗨，露西，有什麼我能幫妳的嗎？

露　　西：我們有個還在進行中的訂單，貨款已經電匯了，但還沒接到任何出貨的消息。

派翠西亞：有，我記得上星期一妳有發訂單給我。讓我查一下訂單狀況…嗯…我們是還沒出貨，那是因為還沒收到匯款喔。

露　　西：真的嗎？我們上星期就匯款了耶！

派翠西亞：那這樣我了解了，請讓我跟妳說明一下狀況。

露　　西：請說。

派翠西亞：通常電匯之後要等個兩個星期，才會確實入到我們的帳戶。

露　　西：真的嗎？可是我們真的急著要貨耶！妳能今天就直接幫我們出貨嗎？

派翠西亞：抱歉，我們公司的政策規定是要收到貨款才能出貨。我會跟我們的銀行追一下妳們這筆電匯，等一入帳後，就立刻安排出貨。

露　　西：好的！請務必幫我們留意一下這件事。

派翠西亞：沒問題，那麼，等我跟銀行談過之後，我再回妳個電話。

露　　西：謝謝妳。

電話英文短句—好說、說好、說得好！

· What can I do for you?　有什麼我能幫妳的嗎？
· Let me check its status.　讓我查一下狀況。
· I'm listening.　請說。
· Please do keep an eye on it for us　請務必幫我們留意一下這件事。

國貿知識補給站 **付款條件與方式**

在國貿操作實務中，雖然花在付款條件討論上的時間通常沒有比產品詢問、下單、催出貨來得多，但錢這事可是一板一眼的，一開始就應把條件談清，把相關的狀況問個明白，才不會因為付款一事而延誤了出貨或其他時程上的安排。現在，就讓我們來了解一下，看看各種付款條件究竟是如何讓貨款從買方的荷包，「錢」進到賣方那兒去！

➤ **信用狀／Letter of Credit (L/C)**

買方向銀行申請開立信用狀，將這張附有條件的付款保證文件，開立給賣方，銀行向賣方承諾，若賣方能履行所規定的條件，並提示出貨文件，則可擁有開狀銀行的付款擔保。

➤ **付款交單／Document against Payment (D/P)**

賣方在貨物裝運後，開出匯票，連同出貨文件，委託銀行交給進口地的代收銀行，請其代為向買方收取貨款，買方須付清貨款後，才能取得出貨文件，辦理提貨清關。

➤ **承兌交單／Document against Acceptance (D/A)**

承兌交單與上述付款交單的程序相同，唯一不同之處在於代收銀行僅須買方在匯票上承兌，即可取得出貨文件，辦理提貨清關，等到規定的付款期限到時，再行付款。

➤ **憑單據付款／Cash against Documents (CAD)**

賣方在貨物裝運後，將出貨文件於出口地交給買方或其代理人，即可自買方或代理人處收取貨款。

➤ **貨到付款／Cash on Delivery (COD)**

賣方在貨物裝運後，將出貨文件交給買方辦理提貨清關後，則可向買方收取貨款。

➤ 預付貨款／Cash in Advance (CIA)、Payment in Advance、
Prepayment

買方預先支付貨款給賣方之後，賣方即可辦理出貨。

在信用狀、付款交單、承兌交單這幾種付款條件中，銀行這一個中介角色皆有收取、轉交出貨文件或／及匯票的任務。對其他的付款條件來說，銀行就無此責任了，有的只是處理買方電匯或開支票的要求。而最快速與無須跑銀行的付款條件，就是信用卡付款、線上付款了。

我們先來說說電匯。電匯除了要知道收款方的銀行戶名、帳號、銀行國際代碼等資料之外，還要先問清楚賣方配合的銀行會扣多少手續費，這樣才不會有匯款短收的問題，也才不會有那種賣方要求補匯那小小差額的情況出現。

刷卡是最快的付款方式了！若是買方要求刷卡付款，而賣方也可接受，那就會有下列這幾種不同的接受、通知方式：

➤ 賣方若可接受以 e-mail 通知信用卡明細，那麼買方大可就在 e-mail 中寫出持卡人的卡號、姓名、效期、卡片背面的三碼驗證碼，讓賣方依之扣款。

➤ 賣方要求買方填寫一份「信用卡授權書」，除了上一點所說的信用卡明細之外，有的授權書中還會要求其他關於買方發票地址、出貨地址、訂單內容與金額等資訊，請信用卡持有人填寫，簽名之後回傳。

➤ 賣方基於安全考量，不想要買方以 e-mail 通知信用卡明細，因此會要求買方改以傳真回覆。

➤ 賣方要求買方要在其網站上線上下單，才能用信用卡在線上付款。

➤ 賣方會發來線上刷卡系統的 Invoice，例如 PayPal 的 Invoice，讓買

方點入連結，進行付款。

那麼這些付款條件，從談保證、說承諾的信用狀，到動動滑鼠就立馬搞定的線上信用卡付款都有，我們不見得每一種都碰得到，但請讓我們自己有「知」的本事，等到真遇到時，自然就可自在地依著我們已知的知識與規範，循著賣方的指示與提示，輕鬆擺平付款事！

PART
1

PART
2

PART
3

PART
4

PART
5

PART
6

PART
7

付款條件說明信函 範例 ・預付指示、接受電匯

Payment Information – for International Customers

Prepayment Instructions

1. Prepayment is required on all international shipments.
2. Once our bank confirms that your wire transfer was received, shipment will be made on the next possible Friday. If the product is out of stock, we will inform you the expected shipping date.
3. The customers are responsible for all import duties, taxes or other special charges (including return freight if shipment is refused by customs or the customer.)
4. Returns of international shipments will NOT be accepted. No refunds will be made.

Acceptable Payment Method

* Wire Transfer
  Bank: The Bank of New York Mellon
      Address: 1 Wall Street, New York, NY 10286, USA
      SWIFT: IRVTUS3N
      ABA: 021000018
      Account #: 8900123428
      For: ALFA BIOMEDICAL COMPANY

Reference: In the "Reference Field", please include our invoice or pro forma invoice number for your order.
* All wire transfer charges should be paid by Sender. Any money shortages in the amount deposited into our account compared to our invoice total will be invoiced separately.

付款條件說明信函 範例 中文

付款資訊 –適用海外客戶

預付指示

1.所有海外出貨皆須預先付款。

2.當我們的銀行確認收到您的匯款後，我們就會盡可能在接下來的星期五安排出貨，若產品沒現貨，我們會再通知您預計的出貨日期。

3.客戶須自行負擔所有的進口關稅、稅負或其他特別費用（包括若貨物被海關或客戶退回所產生的退運運費。）

4.海外出貨不接受退貨，不接受退款要求。

可接受的付款方式

* 電匯

銀行：紐約梅隆銀行

地址：美國紐約曼哈頓華爾街 1 號（10286）

銀行國際代碼：IRVTUS3N

美國銀行協會代碼：021000018

帳號：8900123428

帳戶名：阿爾法生醫公司

參考：在「參考欄」中，請列出您訂單所對應之我方發票號碼或形式發票號碼。

* 所有的電匯費用應由匯款人支付，若存入我方帳戶的金額少於發票總額，則將另行開立發票收取。

信用卡授權書格式 範例

CREDIT CARD AUTHORIZATION FORM

Customer/Company: Winner Corp.	Billing Address: No. 155, JenJen Street, SanChung District, New Taipei City, Taiwan
Customer PO No.: TW100	
Card Holder's Full Name: MeiMei Liu	
Phone No.: 886229721000	Shipping Address: No. 155, JenJen Street, SanChung District, New Taipei City, Taiwan
Fax No.: 886229721001	
Card Type: MASTER Card	
Credit Card No.: xxxxxxxxxxxx1168	
Credit Card Expiration Date:03/2018	Requested Delivery Date: 03/20/2015
3 Digit Security Code on Back of Card: 132	Date Order Placed: 03/10/2015

Subtotal:	US$ 1,000
Estimated Special Handling Fee, Freight, and Sales Tax (If Applicable):	US$ 150
Total Order Invoice Amount/Amount to be Charged to Credit Card:	US$ 1,150

Item	Part #	Qty.	Unit Pirce	Description	Total
1	101	10	US$ 100	101 Alarm Clock	US$ 1,000

Card Holder authorizes payment and agrees to our standard terms of sale.

Signature of Card Holder: *(Please sign here)*

信用卡授權書格式 範例 中文

信用卡授權書格式

客戶／公司： 威拿公司	發票地址： 臺灣新北市三重區人人街 155 號
客戶訂購單單號：TW100	
信用卡持有人全名： MeiMei Liu	
電話號碼：886229721000	出貨地址： 臺灣新北市三重區人人街 155 號
傳真號碼：886229721001	
卡別： MASTER Card	
卡號： xxxxxxxxxxxx1168	
信用卡效期：03/2018	要求的出貨日： 03/20/2015
信用卡背後三碼驗證碼：132	下單日： 03/10/2015

小計：	US$ 1,000
特殊手續費、運費與銷售稅預估（如適用）：	US$ 150
總訂單發票金額／待扣信用卡款項金額：	US$ 1,150

項次	貨號	數量	單價	品名	總計
1	101	10	US$ 100	101 鬧鐘	US$ 1,000

信用卡持有人授權付款，並同意我方銷售標準條款。
信用卡持有人簽名： (請在此簽名)

Part 4 · Unit 2 付款

PART
1

PART
2

PART
3

PART
4

PART
5

PART
6

PART
7

 《Dear Amy》時間

Dear Amy，

我大學唸的是財管系，但對於那些經濟啊會計的，實在沒什麼熱情，所以畢業後，我就選了英文秘書的工作來做，可是，現在的我居然需要開始核對一堆 Invoices 的帳，還要看長長的應付未付帳款清單，我都快被這些呆帳打得腦子變呆了哩…請問 Amy，像這類付款的事和核帳的工作，有沒有辦法不要讓它那麼惱人啊？平時要怎麼做才不會以後有呆帳逼得我核對呢？請開示，謝謝。

想要不食人間煙火卻陷在舊帳錢味裡的 Kathy

Dear Kathy，

我也很想過著不食人間煙火的清雅生活，可是我太愛人間煙火料理出來的佳餚美食啊！恭喜妳已經往「通才」的路上邁進了！妳的角色是與國外連絡的窗口，因此，舉凡所有業務、技術、會計事，只要有需要跟國外討論，就都會是妳的事，而當這些原先我們不懂但我們得硬著頭皮去做的事，都是讓我們懂更多、變通才的好機會呢！先來說說 什麼會變呆帳？什麼得要我們耗神追查往事呢？那都是因為當初並沒有馬上將帳務理清，才會遺留在帳上清不掉。那現在到了我們手上，要怎麼追查呢？請在國外原廠來的帳款清單表上加註，逐條說明，看我們手上有沒有該筆 Invoice，沒有的話就請原廠補來，若 Invoice 已付款，就請附上當初的銀行匯款證明，若 Invoice 有疑義，像是賠償出貨不該收費，那就請將當初客訴、賠償的往來信件附上。在處理這樣的帳務事時，有一個重點一定要注意：呆帳之所以愈待愈久，就是原廠發現了、問了，我們查了、回了之後，原廠又沒下文了，這時請切記卯起來追、定時追，一定要追到原廠回覆、處理了，才能確保那些呆帳不會三個月後又回到妳的手上要妳查。所以，對到付款與帳務事，請記得兩大處理原則：新 Invoice 有問題，就要當下馬上反應，要求修改，而在追查舊 Invoice 資料時，查了就盯著原廠解決，這樣我們才能高枕無憂，悠哉地邊工作邊享用人間煙火美味點心呢！

單字片語說分明

- **credit** [ˋkrɛdɪt]

n. 存入帳戶的錢 an amount of money that you add to an account; cf. an amount of money that you take out of an account is a debit.

例 If payment has been sent, there would be a credit to use in your account in the future.

若是您已付款，您的帳戶上就會有一筆存款，可供將來使用。

n. 賒帳 an arrangement to receive goods from a shop or money from a bank and pay for it later

→ credit card 信用卡: a small plastic card that you use to buy things now and pay for them later

→ credit note / memo: 貸項通知單、貸項憑單

A credit note can be issued to correct a mistake if the invoice has been overstated or to reimburse the buyer completely if the goods have been returned.

若發票金額有高列的情況，或是來貨已被退回，可全額退款時，則可開立貸項通知單，以用來更正錯誤。

常見搭配詞

credit limit 信用額度　credit rating 信用評等　interest-free credit 免息賒帳

- **transfer** [trænsfɚ]

n. 轉讓；轉帳 the act of moving somebody/something from one place, group or job to another

例 Payments can be made by bank transfer, credit card or mailed check.

可透過銀行轉帳、信用卡或郵寄支票來付款。

其他字義：調職；運送

transfer [trænsˋfɚ] Ⓥ

常見搭配詞：名詞＋ transfer 名詞

credit transfer 銀行轉帳　　　　　technology transfer 技術移轉

常見搭配詞：動詞 transfer ＋ 名詞
transfer power/ authority/ responsibility 移交權力／職權／職責

- -

· issue [ˋɪʃʊ]

v. 發佈；核發 to announce something, or to give something to people officially

例 We have put the request to our Finance team to issue a credit memo for the duplicated Cat# 438508 on Inv # SI159609.
對於號碼 SI159609 這張發票裡所重覆列出的型號 438508，我們已要求我們的財務組核發一張貸項通知單。

其他字義：（使）流出；由…產生；出版

issue n. 問題；爭議；發行（物）；（期刊的）期、號

常見搭配詞：名詞 issue

contentious/ controversial/ thorny issue 有爭議／引起爭論／棘手的問題

raise an issue 提出（須討論）的問題

confuse/ cloud/ fudge the issue 把事件處理得很亂

not be an issue 不成問題

- -

· advance [ədˋvæns]

adj. 預先的 done, obtained, or announced before a particular time or event

例 A purchase order or a full advance payment is required for ordering. For custom synthesis projects, a 50% upfront payment is needed to initiate the project.
下單時必須要提出訂購單，或是預先全額匯款，對於訂製合成的專案，待先行支付 50%的貨款之後，我們才會處理該專案。

→ in advance 事先；預先

例 For international customers, we require payment in advance before products can be shipped.
對於海外客戶，我們要求出貨前預付貨款。

其他字義：n. .進展；預付款；（尤指軍隊的）前進；（尤指不受歡迎的）勾引、挑逗

advanced [əd`vænst] adj 先進的

advanced degree （學士以上的）高等學位

advancement [əd`vænsmənt] n 升遷；晉升；（社會、科學、人類知識的）進展；促進

· **installment** [ɪn`stɔlmənt]

n. 分期付款 one of several payments that an amount is divided into, so that you do not have to pay the whole amount at one time

= instalment （英式拼法）

例 You can pay in installments of any amounts that are convenient for you. However, if it's a last minute booking, full payment is due immediately.

您可依您方便，支付任何金額分配的分期付款，但若是到最後一刻才下單，則須立即支付全額。

其他字義：安裝；就任；安頓

install [ɪn`stɔl] V

· **applicant** [`æpləkənt]

n. 申請人 someone who applies for something, such as a job or a loan of money

例 Please send the completed authorized application form and a blank copy of your official company letter head to us.

請寄給我們您所填寫完的授權申請表，以及一份印有您公司信頭的正式信紙影本。

apply [ə`plaɪ] V                                applicable [`æplɪkəbl] adj 適宜的

applied [ə`plaɪd] adj （學科）應用的；實用的  appliance [ə`plaɪəns] n 家用電器

application [ˌæplə`keʃən] n 申請；用途；應用軟體

常見搭配詞

make/ submit/ put in an application 提出申請

grant/ approve an application 批准申請

- beneficiary [ˌbɛnəˈfɪʃərɪ]

n. 受益人 someone who applies for something, such as a job or a loan of money

例 Who is the main beneficiary of this deal?

誰是這筆交易的主要受益人呢？

benefit [bɛnəfɪt] n 好處；津貼；福利

beneficial [ˌbɛnəˈfɪʃəl] adj 有益的

beneficent [bɪˈnɛfəsənt] adj 行善的；慈善的

benevolent [bəˈnɛvələnt] adj 樂善好施的；行善的；善意的

常見搭配詞

mutual beneficial 互利的

get/ gain/ receive a benefit 得到利益

reap a benefit 受益

- signature [ˈsɪgnətʃɚ]

n. 簽名 a person's name written in a special way by that person

例 I've noted that there are two identical signatures on the "Material Transfer Agreement". This will not be accepted by our Authority.

我注意到在「材料移轉合約」上有兩處簽名是一樣的，我們的主管機關對此是不會接受的。

sign [saɪn] v

常見搭配詞

signed and sealed （協定）已經生效的　sign away 簽字放棄（或讓與）

sign for 簽收（包裹或信件）　　　sign in 登記；簽到

sign on 登入（電腦）；僱用　　　electronic signature 電子簽名

Part 5
問題處理篇

Unit 01

催貨與催款

單字片語一家親

問貨況

缺貨待出訂單　backorder

最新情況　update

狀況；情形　status

進一步的　further

階段　stage

提醒　reminder

需貨恐急

急迫　urgency

簽訂合約　enter into a contract

規定；約定　stipulate

最後限期　deadline

交貨延遲　delivery delay

違約　breach of contract

處以罰金　fine

罰款　penalty

許一個快快來貨的願

加快執行過程　expedite process

縮短等待時間

shorten the waiting time

在限期前完成（交貨）

meet the deadline

鬆開（放行）；發佈　release

催款來著

提醒　reminder

逾期通知　overdue notice

錢呢？

尚未償付的 outstanding, pending

逾期的　overdue, past due

逾期剩餘款　overdue balance

拖欠帳目　delinquent account

應收帳款　accounts receivable

再不付的話…

信用凍結　put on credit hold

帳戶凍結　account suspension

終止帳戶　account termination

對不住…

抱歉　apology

遺憾；感到抱歉　regret

晚回覆　late response

不便　inconvenience

影響　impact

導致；引起 cause

🔑 句型

句型 1 杳無音訊？

It's been over + 期間 + since we heard the news about sth.

例 It's been over half a month <u>since we heard the news about</u> the progress of your export permit application.

我們已有半個多月沒聽到有關您出口許可申請進度的消息了。

句型 2 有消息否？

Have you got any further update on sth?

例 <u>Have you got any further update on</u> the release of the new product?

關於新產品推出這件事，您有任何進一步的消息嗎？

句型 3 記得我們有約

We expect our order can be shipped no later than + 日期.

例 <u>We expect our order can be shipped no later than</u> 3 working days after the order is placed.

我們希望我們的訂單在下單後三個工作天之內就可出貨。

句型 4 不是好消息…

We regret to inform you that…

例 <u>We regret to inform you that</u> the following items have been delayed.

很遺憾我們得通知您下列這些品項有延遲的狀況。

句型 5 抱歉

Please accept our apology for sth.

例 Please accept our apology for any inconvenience that has been caused.

對於造成您的任何不便，請接受我們的道歉。

✉ 催貨 E-mail | **Request for Shipment**

催貨 E-mail 範例 ・合約中設有交貨限期，遲交得開罰

Dear Gary,

It's been over two week since we heard the news about the status of our following backorder. According to the information we were provided, the product should be released by the end of this month. Have you got any further update on this matter? Please advise.

PO No.	Cat. No.	Qty.	Product Description
1031217	AC-33F	10	Microscope Custom S-1 System

As you know, we have entered into a contract with our customer for this order. The delivery deadline stipulated in the contract is July 3rd. Therefore, we expect our order can be shipped no later than June 25th. If we fail to meet the specified deadline, we'll be charged a penalty for delivery delay. Please do help expedite the production process and let us know when the product can be shipped to us at the soonest. Thanks.

Sincerely,

Vicky Tseng
Central Instrument Co., Ltd.

單字 ShabuShabu 一小補小補！
☑ microscope [`maɪkrəˌskop] (n.)
   顯微鏡
☑ specify [`spɛsəˌfaɪ] (v.)　指明、規定
☑ instrument [`ɪnstrəmənt] (n.)　儀器

催貨 E-mail 範例　中文

蓋瑞，您好，

　　對於下列這筆未出貨訂單，我們已有兩個多星期沒聽到進一步的消息了。按照先前您所給我們的訊息，這項產品應可在這個月底前供貨，請問您有任何新的消息嗎？還請告知了。

訂單單號	型號	數量	品名
1031217	AC-33F	10	特製 S-1 系統顯微鏡

　　您知道我們此單已與客戶簽了合約，合約中規定的交貨期限為七月三日，因此，我們希望最晚可在六月二十五日前出貨。如果我們沒能在此限期前將貨交給客戶，將會因交貨延遲而被罰款。請務必幫忙加速生產處理程序，也請告知此產品最快何時可出貨，謝謝。

曾薇琪
中央儀器有限公司

Check 好句

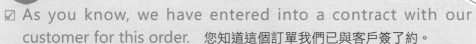

☑ As you know, we have entered into a contract with our customer for this order.　您知道這個訂單我們已與客戶簽了約。
☑ If we fail to meet the specified deadline, we'll be charged a penalty.　如果我們沒能在此限期前將貨交給客戶，就會被罰款。

回覆催貨 E-mail 範例 ・貨還沒好、會查查能否早些供貨

Hi Vicky,

Please accept our apology for not keeping you updated on your order status.

We regret to inform you that the status of your backorder remains unchanged. Our Manufacturing Dept. is still in process of this item. As it is custom made, the lead time is 3 full weeks.

As far as I know, the production currently is still scheduled and ready for dispatch in the end of this month, depending on the fact that the product will leave the QC stage successfully. Meanwhile, I've also contacted our responsible manager to ask if we will be able to supply the product before the projected date. As soon as I have any updates, we will inform you.

We appreciate your understanding and patience.

Best regards,

Gary Jackson
Kingdom Technology

單字 ShabuShabu 一小補小補！
☑ depend on　視⋯而定
☑ meanwhile [`min͵hwaɪl] (adv.)　同時
☑ project [prə`dʒɛkt] (v.)　預計

回覆催貨 E-mail 範例 中文

薇琪，您好，

　　抱歉沒有跟您更新訂單的狀況。

　　很遺憾，您此份未出貨訂單的狀況跟先前還是一樣，這項產品仍在製造部門生產中，因它是訂製產品，所以接單後要足足三個星期才能供貨。

　　就我所知，此產品目前仍是預計在這個月底才能完成生產，才能出貨，不過前提是它可成功通過品管檢測階段。同時，我也有跟負責此產品的經理詢問過，看是不是有可能在預定的日期前供貨，等我一有新的消息，就會馬上告訴您。

　　謝謝您的諒解與耐心等候。

祝好

蓋瑞 · 傑克森
王國科技

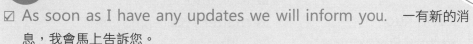

必 Check 好句

☑ As soon as I have any updates we will inform you.　一有新的消息，我會馬上告訴您。

☑ We appreciate your understanding and patience.　謝謝您的諒解與耐心等候。

催款 E-mail | Request for Payment

催款 E-mail 範例 · 貨款逾期 30 天，不回或不付就要凍結信用了

Dear Cathy,

Attached to this e-mail you will find our reminder for your outstanding account. If you are not the person in charge, please be so kind and forward this e-mail to your Accounting Department.

This e-mail is to inform you that your payment of £5500.00, against Invoice # 0925, is now 30 days overdue. Please look into this matter and make the payment asap so as to maintain good business relations between our two companies.

If you do not respond to this e-mail or make the payment within 10 days, we regret to inform you that we will have to place your company on credit hold due to the past-due invoice. Please let me know if you have any questions. We look forward to your reply.

Regards,

Kevin Kuo
Victory Electronics

單字 ShabuShabu 一小補小補！
☑ in charge 負責
☑ look into 調查
☑ matter [ˋmætɚ] (n.) 事情
☑ victory [ˋvɪktərɪ] (n.) 勝利
☑ electronics [ɪlɛkˋtrɑnɪks] 電子

催款 E-mail 範例 中文

PART
1

PART
2

PART
3

PART
4

PART
5

PART
6

PART
7

凱西，您好，

在此提醒您，您的帳上還有未償付帳款如附。若是您不負責處理此事，還請您將此 e-mail 轉給會計部門。

我們這一封 e-mail 是要告訴您，您發票號碼 0925 的這筆 5500 英鎊的款項，已經逾期三十天了，請查查此事，並儘快付清貨款，以維持我們雙方良好的商務關係。

若是您未在十天內回覆此事或付清貨款，很遺憾地我們就必須因這筆逾期貨款而凍結您公司的信用。若您有任何問題，請告知。等候您的回覆。

祝好

郭凱文
勝利電子

✓ Check 好句

☑ If you are not the person in charge, please be so kind and forward this e-mail to your Accounting Department. 若是您不負責處理此事，還請您將此 e-mail 轉給會計部門。

☑ Please look into this matter and make the payment asap. 請查查此事，並請儘快付清貨款。

回覆催款 E-mail 範例 · 不小心漏掉了，馬上補匯

Dear Kevin,

Please accept our apology for failing to make payment on time. The Invoice # 0925 was missed to be included in our Account Department's wire transfer list of last month by oversight and we're not aware of this until receiving your this overdue notice.

We'll wire the entire balance of £5500.00 first thing tomorrow morning and e-mail the remittance advice to you. If you do not receive it within a week, please do remember to notify us.

We also received an e-mail directly from your Accounting Department saying that our account has been placed on credit hold. As stated above, we'll pay for the overdue Invoice tomorrow and so please help have the hold released.

Sorry again for any inconvenience caused to you.

Sincerely,

Cathy Bullock
Eternity Electronics

單字 ShabuShabu 一小補小補！
☑ oversight [`ovɚˏsaɪt] (n.) 疏忽
☑ balance [`bæləns] (n.) 餘額
☑ inconvenience [ˏɪnkən`vinjəns] (n.) 不便
☑ eternity [ɪ`tɝˏnətɪ] (n.) 永恆

回覆催款 E-mail 範例 **中文**

凱文，您好，

　　很抱歉我們沒有準時支付貨款，我們會計部門在處理上個月發票的匯款作業時，不小心將這張號碼為 0925 的發票遺漏了，直到收到您的逾期通知，我們才知道漏了這張發票。

　　我們明天一早就會去匯這整筆 5500 英鎊的未付貨款，並 e-mail 匯款通知給您，若您在一個星期內沒有收到這筆匯款，還請記得通知我們一聲。

　　我們另外也有收到您會計部門直接發來的 e-mail，說已對我們的帳戶做了信用凍結。如前所述，我們明天會付清逾期發票，所以就請您幫忙解除此信用凍結。

　　對於造成您的不便，在此再次跟您說聲抱歉。

謹上

凱西・布拉克
永恆電子

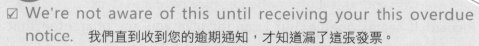

必 Check 好句

☑ We're not aware of this until receiving your this overdue notice. 我們直到收到您的逾期通知，才知道漏了這張發票。

☑ We'll wire the entire balance of £5500.00 first thing tomorrow morning. 我們明天一早就會去匯這整筆英鎊 5500 的未付貨款。

☎ 電話對話

電話催貨 範例

人物介紹

Rose

蘿絲，業績不錯，但常有迷糊事，考驗大家的危機處理能力！

Richard

理查，做事有彈性，只要覺得有可能就願意嘗試，以求解決問題。

Rose: Hello. Is this Richard?

Richard: I's Richard speaking. May I ask who's calling?

Rose: My name is Rose Lo from Grand Engineering. I placed an order of 10 controllers to you the day before yesterday.

Richard: Yes, I'm aware of that order. What can I do for you, Rose?

Rose: Well, I'd like to know whether you could ship for us this Friday?

Richard: This Friday? Impossible! The lead time of controllers is 2 weeks ... And I did notify you of this, didn't I?

Rose: Yes, you did. But for the date of delivery stipulated on the contract with our customer, I wrongly thought it's the date of shipping.

Richard: That's too bad... Unfortunately, the controllers will not be released until late next week at the soonest.

Rose: Is there any other way for us to get the product? Could you check with your distributors whether they have it in stock?

Richard: It seems doable, but I cannot give you any guarantee that it works because we haven't done this before.

Rose: I understand, but please still give it a try. We do need the product urgently.

Richard: OK. I'll do that and let you know what we got!

電話催貨　範例 中文

蘿絲：你好，是理查嗎？

理查：我是，請問是哪位呢？

蘿絲：我是格蘭德工程的羅蘿絲，我在前天下了一份 10 個控制器的訂單給你。

理查：有的，我知道那份訂單，請問有什麼我能為妳做的嗎，蘿絲？

蘿絲：是這樣的，我想要問問你有沒有辦法在這個星期五替我們出貨？

理查：這個星期五？不可能的啊！控制器的出貨準備期是兩個星期…我先前有告訴妳這件事，沒有嗎？

蘿絲：沒錯，你是有說，不過在我們跟客戶簽的合約中，我誤把交貨日看成出貨日了。

理查：那真糟糕…可惜控制器最快也是要到下星期週末前才可能供貨。

蘿絲：有其他辦法可讓我們拿到貨嗎？你能跟你們的經銷商問問有沒有備著存貨嗎？

理查：這好像可試試，不過，因為我們之前都沒這樣做過，我沒辦法保證行不行得通。

蘿絲：我瞭解，但請還是試試，我們真的急著要這批貨。

理查：好的，我會去試，有什麼結果會再告訴妳囉！

電話英文短句─好說、說好、說得好！

・May I ask who's calling?　請問是哪位呢？

・That's too bad.　真糟糕。

・Please still give it a try.　請還是試試。

國貿知識補給站 「催」的理與禮

當事情到了要「催」的地步，八成都是有人落了進度、做了沒做全、說了沒做到，硬是讓人等得望穿秋水，等到心急如焚。這時候該怎麼辦呢？一個勁兒給它催去，外加責怪與威脅？喔不？我們可是講道理的文明人，怎會如此無禮呢？注意！兩個關鍵字出現了，那就是「理」與「禮」！我們可以盡量催、盡量說，但請記得只說「理」，而且要合「禮」。現在，就讓我們一起從「催」的內容（以催貨為例）與時間，來看看守理說理、行禮如儀該有的應對態度吧！

「催」的理：大風「催」，「催」什麼？

要催訂單，若要催得清楚、一催就中，可就得好好說明下列這幾點呢！

1. 破題點出哪份訂單、哪件事情，寫個清楚

為了要讓收信的對方能夠馬上接上你所要談論的訂單與事情，例如在說哪一筆訂單時，除了說你們自己下單的單號，應也一併提供訂單日期與原廠的訂單確認單單號，讓原廠一下子就能清楚地知道所談何事。

2. 寫明原預訂出貨日、需貨急切的狀況

催促的 e-mail 裡，除了「急、急、急」這急到口吃的訊息之外，請給原廠一個急的事實與情境，讓他們也能想像，好能有同步且急切的應對態度。

3. 說出確切的要求

對於要原廠急著處理、快些回覆的事，請記得告訴原廠「哪一天」之前一定要他們有回應動作（如答覆、出貨），以及要求原廠做到、

做足「哪些事情」，這樣才能讓原廠知道你的時程要求，也才更能確保原廠做事做足整套，不會讓你好不容易催了對方也回覆了，卻發現既不全，也不夠，這樣就還需要再發 e-mail、再催人、又再催事。

4. 說明後果與後患

　　跟國外催一件事時，事後常會發現其實你這邊急得一頭熱，國外卻還在納涼，這就是資訊不對等之下可能發生的情況，也是要求的人得自行承擔的後果。若是你有急切的理由，知道此次任務沒達成會有什麼樣的後果與損失，那就請你確實說明。事件的相關人都有知的權利，也有「全知」的必要，因此，要求時請以充分的資訊為基礎，這樣才能讓對方有最為完整與完全的處理。

「催與被催」的禮：早催、早回！

1. 提早催

　　催促這事的首要禮儀就是：早些提醒、早些催！不要等迫在眉睫時才催，讓雙方都急翻了天！像是對於交貨日期有畫押的重要訂單，若務求準時交貨，不能有任何一點兒閃失發生，這時就請務必提早提醒原廠預定的出貨日，例如一個星期前詢問原廠，順道了解貨準備好了沒？有沒有發生什麼新的狀況？做了這個動作，可提醒原廠不要忘了約定的時間，不要漏了處理約定的事。而且，主動即早詢問，則可免去遇到讓人措手不及的突發狀況。

2. 馬上回

　　催貨的人會是被催款的人，催款的人也就是被催貨的人，所以在國貿這事上，最有機會讓人修鍊將心比心的好性情了！話說事情到了催的地步，就是有急著解決的必要，所以，收到催促 e-mail 時，最有禮、最恰當的回應方式，就是快快回，當天回！實際上，對於這類催促，收到後多半得跟生產部門或會計部門查詢，才會有個答案，所以

處理的人常會因暫時還沒有答案而就先不回覆，這不是個適當的處理方法，當然也是讓人覺得處理的人置之不理的無禮對待。其實秉持著待人接物的道理來處理就對了：有人問了你一個問題，你就該回答，就算回答當下沒有別的資訊可提供，也至少可以告訴對方你會怎麼處理，預計何時可再給對方回覆，這樣才是有禮又負責的做人處事之道啊！

《Dear Amy》時間

Dear Amy，

我工作的公司代理了很多家進口零件原廠，所以我一天到晚就忙著下單、催貨，也忙著被催，被追著付款—這點讓我有些頭痛，因為我每天都會收到好幾封催款 e-mail，有系統自動催款信，有國外一到發票日後 30 天就來催款的信，可是我們明明都已月結給國外，他們卻還這樣常發催款信，催得我開始有點煩躁了，有時還叫我們簽名，好複雜喔！請問這事有簡單版的處理法則嗎？

沒人追求但一直被追帳的 Luisa

Dear Luisa，

看到妳的署名，我覺得我們應該要聊追求事耶！不過，妳說想知道的是簡單版的對策，那我就確定這只能聊追帳事了，因為感情事就算表面看來不複雜，內心也總是會有糾結啊！好了，讓我們認命地來看看被催款的事吧！

1. 系統自動催款信：只要妳確定國外也同意每月月結一次的方式，那麼這就是難得的妳也可選擇「已讀不回」的信，不會失禮，不會有人追殺過來。

2. 國外廠商設定付款天數所發來的催款信：妳們已談定月結，所以這也是提醒性質，那就請回覆妳們都是每個月哪一天付款即可。

3. Statement／對帳單：原廠將「Accounts Receivable／應收帳款」列表呈現，當妳在回覆這樣的對帳單時，就請說清楚哪些帳已付、哪些未付、哪筆帳有問題，哪筆 Invoice 並沒有收到。

4. Confirmation of Balance／帳目餘額證明書：妳說到要簽名的催款信就是這個了，它不是用來催款的，是會計查帳的查核文件，要跟妳確認到「某日」為止，妳們帳戶上的應付未付 Invoices 資料是否正確。這就請核對後簽回即可。

最後，提醒一下，對於帳務上的這些提醒與催促，請記得來一封解決一封，也請會計部務必配合，快快回覆、快快處理，才不會讓帳就此一直擱著，愈積愈多，也才不會因付款、回覆的延遲，壞了與國外建立的和諧關係喔！

🔍 單字片語說分明

• update [`ʌpdet]

n. 最新情況 a report or broadcast containing all the latest news or information

例 Unfortunately, there is no further update and no firm shipping date has been set.

很遺憾的是，我們尚無進一步更新的消息可提供, 可供貨日期也還沒確定下來。

update [ʌp`det]

v. 更新 to add the most recent information to something such as a book, document, or list

例 We shall update you as soon as we have news of release.

一有供貨相關的消息，我們就會立刻跟您更新說明。

• status [`stetəs]

n. 狀況；情形 the level of importance or progress in a particular situation or discussion

例 Could you explain a little bit more about the current status of market structure?

你能稍微解釋一下市場結構的現況嗎？

常見搭配詞

social status 社會地位   professional status 專業地位

high/ privilege status 崇高的地位   status quo 現狀

status symbol 身份地位的象徵   status bar （電腦螢幕下方的）狀態列

- reminder [rɪ`maɪndɚ]

  n. 提醒 something that reminds you of something that happened in the past

  例 Just a reminder that our office will be closed on Monday for our Presidents Day Holiday.

  在此提醒一下，星期一為總統日假期，我們公司會放假一天。

  remind [rɪ`maɪnd] Ⓥ

- urgency [`ɝdʒənsɪ]

  n. 緊急；迫切 the need to deal with something quickly

  例 I will notify Production Department about the urgency of your order.

  我會通知生產部門您的訂單急要貨。

  urgent [`ɝdʒənt] adj

  常見搭配詞

  | matter of urgency 緊急的事 | sense of urgency 急迫感 |
  | in urgent need 急需 | urgent message/ appeal/ call 緊急呼籲 |

- stipulate [`stɪpjə‚let]

  v. 約定；規定 to say what is allowed or what is necessary

  例 We will be stipulating a minimum order quantity for X-series products.

  對於 X 系列產品，我們將會規定最小訂單訂購量。

  stipulation [‚stɪpjə`leʃən] Ⓝ

- deadline [`dɛd‚laɪn]

  n. 最後期限 a specific time or date by which you have to do something

  例 If you would like to have your orders supplied within the end of this year, the deadline for receiving your orders will be December 12th.

PART 1
PART 2
PART 3
PART 4
PART 5
PART 6
PART 7

若您的訂單想要在今年年底前出貨的話，最晚請於十二月十二日下單給我們。

常見搭配詞

meet a deadline 按時完成　　　　miss a deadline 沒有按時完成
set a deadline 定出最後期限

• breach [britʃ]

n. 違反 a failure to do something that you have promised to do or that people expect you to do

例 If one of the parties to a contract fails to perform as required by the contract, this may constitute a breach of contract.

如果合約中有一方無法依合約要求來執行，即構成了違約行為。

breach [britʃ] Ⓥ

常見搭配詞

breach of confidentiality agreement 違反保密協定

breach of duty 失職　　　　breach of trust 背信

• penalty [`pɛnḷtɪ]

n. 處罰 a punishment imposed for breaking a rule, law, or contract

例 Could you please elaborate what the penalty is for these kits since they are under contract?

這幾組的貨既然有簽合約，您能不能說明清楚罰則是什麼呢？

penalize [`pinḷ͵aɪz] Ⓥ

常見搭配詞

severe/ heavy/ stiff/ tough penalty 嚴厲的處罰

carry a penalty 判處處罰

· release [rɪ`lis]

v. 鬆開 to stop holding something

例 If the product passes QC, it will be released. If it fails, then they will need to restart production.

如果產品通過品管檢測，就可放行，若沒通過，就要重新生產。

其他字義：釋放；公佈；發行；消除（不良情緒）；免除

release [rɪ`lis] n

· outstanding [`aʊt`stændɪŋ]

adj. 尚未償付的；未解決的 not yet paid, resolved, or dealt with

例 Would you like to pay these outstanding invoices using your credit card, and would you like all future invoices paid likewise?

對這幾張尚未付款的發票，您要用信用卡支付嗎？以後所有的發票也都是要這樣扣款嗎？

其他字義：傑出的

outstandingly [`aʊt`stændɪŋlɪ] adv 非常

· due [dju]

adj. 到期的 if money is due, it is time for it to be paid

例 It is best to give your customer at least a week's notice before the payment is due to be made.

最好是在貨款到期一個星期前，通知你客戶一聲。

其他字義：按照規定的；充分的

常見搭配詞

have due regard 給予充分尊重 | give due consideration 給予充分的考慮
in due course 在適當的時候 | with (all) due respect 恕我直言
due date 期限 | due to 由於

客服與客訴

單字片語一家親

呼叫！請求協助

客戶服務　customer service
電話客服中心　call center
技術支援　technical support
技術協助　technical assistance

要資料來著！

操作手冊　operating manual
操作說明書
operating instructions
規格手冊　specifications manual
使用者手冊　user's manual
技術手冊　technical handbook
產品說明書　product booklet

問題來了

遇到問題　encounter a problem
性能問題　performance problem
困難的　difficult
差異　difference
應用　application
故障　breakdown, malfunction
故障的　out of order
　　　　out of function

要的不只是協助⋯

投訴　complaint

賠償　compensation,
reimbursement
換貨　exchange
替補的貨　replacement
替代品　substitute
退款　refund
折扣；退款　rebate
現金折扣　cash rebate
退貨　return
退貨授權書
Return Authorization

動手解決問題

問題排除　troubleshooting
處理　deal with, cope with,
take care of, manage, tackle
回饋　feedback
損害控管　damage control

就是要找你

技術人員　Technician
技術代表
Technical Representative
產品專員　Product Specialist
故障檢修員　Troubleshooter
工程師　Engineer

句型

句型 1 不同在哪裡？

Please tell me what the difference is between A and B.

例 Please tell me what the difference is between your product and competitor's.

請告訴我您的產品與競爭者的產品有什麼不同的地方。

句型 2 送上資料

Please see sth attached for a complete reference.

例 Please see the specifications manual attached for a complete reference.

請見附件的規格手冊，讓您有個完整的參考。

句型 3 有狀況

We're currently experiencing problems with sth.

例 We're currently experiencing problems with our online login system.

我們的線上登入系統目前有問題。

句型 4 偵探出動！

Our best guess is that…

例 Our best guess is that the competitor will offer an extremely low price to vie for this order this time.

我們認為這家競爭廠商這次會出一個低到不行的價格來搶單。

句型 5 抱歉讓你失望了！

We're sorry to hear that sth is not performing as well as expected.

例 We're sorry to hear that the equipment is not performing as well as expected.

很抱歉這個設備的表現不如預期。

PART 1

PART 2

PART 3

PART 4

PART 5

PART 6

PART 7

國貿英語 溝通術
Master English — Communication for International Trade

客服 E-mail 範例 ・新舊版產品有何不同？

Dear Henry,

We had purchased TBAR Assay Kit, Item number 705000, from you last June. We're now evaluating its new generation products as follows:

TBARS Assay Kit, item number 706002
TBART Assay Kit, item number 707002

Please tell me what the difference is between the old and new versions. Can these new version items cross react with animal proteins? Is there any references citing the use of these new kits? Moreover, do they have different sample preparation procedures? Please reply and e-mail to me its related data for our reference. Thanks.

Regards,

Tina Chen
Advanced Technologies, Inc.

單字 ShabuShabu 一小補小補！
☑ generation [generation] (n.) 一代
☑ react [rɪ`ækt] (v.) 反應
☑ advanced [əd`vænst] (adj.)
　 先進的；進階的

客服 **E-mail** 範例 中文

亨利，您好，

　　我們在六月時有跟您購買了 TBAR 試劑組，型號 705000，目前我們正在評估其下列新一代產品：

　　TBARS 試劑組，型號 706002
　　TBARS 試劑組，型號 707002

　　請告訴我們新版與舊版產品之間有何不同，新版產品可與動物蛋白質交互反應嗎？有任何文獻使用新版產品嗎？此外，新舊版產品的樣品製備程序有無不同呢？請您回覆，並請 e-mail 給我相關的資料，供我們參考，謝謝。

祝好

陳蒂娜
先進科技公司

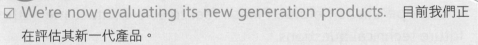

必 Check 好句

☑ We're now evaluating its new generation products.　目前我們正在評估其新一代產品。

☑ Please reply and e-mail to me its related data for our reference.　請您回覆，並請 e-mail 給我相關的資料，供我們參考。

回覆客服 E-mail 範例 · 新舊版產品不同、附上資料

Dear Tina,

Thank you for your e-mail today inquiring our Item number 706002 / TBARS Assay Kit and Item number 707002 / TBAS-1 Assay Kit. Our new TBARS Assay Kits are more sensitive and specific to TXB compared to our old version kit. The new assay kits have not been validated with animal protein samples, but should be suitable according to the kit booklet instructions. Please see the kit booklets attached for a complete reference. I also attached the reference citing the use of this kit with animal samples.

You also asked to know whether they have different sample preparations. The attached kit booklet includes the related information as well. Be aware that the new kits do require more extensive sample preparation than old version kits. They require a purification step. In general, the method of purification is left to the individual investigator.

I hope this is helpful. Please feel free to contact me with future technical questions.

Kind regards,

Henry Park
Plateau Biomedical

單字 ShabuShabu 一小補小補！
☑ sensitive [sensitive] (adj.)   敏感的
☑ specific [spɪˋsɪfɪk] (adj.)   特定的
☑ validate [ˋvæləˌdet] (v.)   證實
☑ purification [ˌpjʊrəfəˋkeʃən] (n.)   純化；淨化

PART 1

PART 2

PART 3

PART 4

PART 5

PART 6

PART 7

回覆客服 E-mail 範例　中文

蒂娜，您好，

　　謝謝您今天發來 e-mail 詢問這兩個產品：型號 706002／TBARS 試劑組及型號 707002／TBAS-1 試劑組。我們的新版產品比舊版的敏感度更高，且特別針對 TXB 而設計。新版試劑組並尚未證實可用在動物蛋白質上，但根據其產品說明書上所列出的資訊來看，應用上應沒有問題。請見附件的試劑說明書，讓您有個完整的參考。另外，我也在此提供此產品用在動物樣品上的參考文獻，請見附件。

　　您也有問到是否它們的樣品製備程序有所不同，在所附的產品說明書中，亦可找到相關的資訊。請注意新版產品的樣品製備所需程序比起舊版更多，因為新版需要純化的步驟，通常，純化的方法都是由研究者自行決定。

　　我希望以上的回覆有幫助到您，往後若有其他技術問題，請儘管與我聯絡。

祝好
亨利・帕克
高原生醫

Check 好句

☑ I also attached the reference citing the use of this kit.　在此我也附上使用此產品的參考文獻。

☑ I hope this is helpful. Please feel free to contact me with future technical questions.　我希望以上的回覆有幫助到您，往後若有其他技術問題，請儘管與我聯絡。

客訴 E-mail 範例 ・產品有問題、要求補出貨、問退運否？

Dear Ted,

　　We ordered M1166 antibody from you early this month. Our Purchase Order no. was 20150202 and the product was shipped against Invoice no. 1218. <u>However, we're currently experiencing problems with the product and need your technical assistance.</u> We have difficulties using this antibody in Western Blotting because it has the problems with a high background and also incorrect band size. Attached please find the testing data for your reference.

　　We have used your this antibody many times before and have not encountered a problem until now. We did carefully follow the operating instructions stated in the product manual. <u>Our best guess is that the antibody shipped this time is inferior in quality.</u>

　　Please let us have your comments on our testing results. On the other hand, since we need this antibody urgently, please ship to us a replacement from another lot this Friday. As to the rest vials of antibody we received, do we need to return to you? Please advise. Thanks.

Best regards,

Ally Huang
Biochem Science, Inc.

單字 ShabuShabu 一小補小補！
☑ blot [blɑt] (v.)　沾上墨漬
☑ band [bænd] (n.)　條紋；帶
☑ inferior [ɪn`fɪrɪɚ] (adj.)
　品質較差的

客訴 E-mail 範例 中文

PART 1
PART 2
PART 3
PART 4
PART 5
PART 6
PART 7

泰德，您好，

　　我們這個月有跟您訂購 M1166 抗體，訂購單單號為 20150202，出貨的發票號碼為 1218，不過，我們現在使用此產品有遇到問題，需要您給予技術協助。我們用此抗體做西方墨點法檢測時有困難，因為背景深，條帶大小也不對。在此附上檢測結果供您參考。

　　您這一項抗體我們已用過好多次了，這還是第一次發生問題。我們確實有仔細照著產品手冊中的操作指示來做，所以我們認為這次所出的抗體品質並不好。

　　請您就我們的檢測結果給予意見，此外，因為我們急著用這一項抗體，請安排星期五從別批批次補出貨給我們。至於我們手上剩下的幾瓶抗體，請問我們需要退回給您嗎？請告知，謝謝。

　　祝好

　　黃艾莉
　　生化科技公司

Check 好句

☑ We have used your this antibody many times before and have not encountered a problem until now.　您這一項抗體我們已用過好多次了，這還是第一次發生問題。

☑ We did carefully follow the operating instructions stated in the product manual.　我們確實有仔細照著產品手冊中的操作指示來做。

回覆客訴 E-mail 範例 · 安排補出貨、要求退運

Dear Ally,

We're sorry to hear that the antibody is not performing as well as expected for your application. There haven't been any other complaints on this antibody, but we will retest the antibody from the same lot of the ones we sent. We expect to be able to send you the retesting results within 2 days.

We can offer you a replacement this Friday to meet your urgency. A Replacement Form is attached to this email, please fill out and e-mail to me. As to the rest vials you received, please return to us (on our FedEX account, of course!). The second attachment is our Return Authorization document which has the instructions and information you will need to ship the item back to us.

The quality of our antibodies is very important to us, as well as the feedback and data we receive. So please do not hesitate to share with us your own experience with our products and let us know immediately any problems you encounter. Thanks.

Best regards,

Ted Alcock
Biosynthesis Technology

單字 ShabuShabu 一小補小補！
☑ urgency [ˋɝdʒənsɪ] (n.) 緊急
☑ fill out 填寫
☑ biosynthesis [ˌbaɪoˋsɪnθəsɪs] (n.) 生物合成

回覆客訴 E-mail 範例 中文

艾莉，您好，

　　聽到您說抗體在應用上表現不如預期，我們感到很抱歉。這一項抗體從未有過任何客訴反應，但我們會從出貨給您的同一批次另取一瓶抗體重新檢測，希望能在兩天內給您我們的檢測報告。

　　我們可以在星期五緊急安排補出貨。附件所附上的是「補出貨表格」，請您填寫後 e-mail 給我。至於您剩下的幾瓶抗體，請您退給我們（當然是走我們的 FedEx 帳號）。第二個附件就是我們的「退貨授權書」文件，裡頭有要退貨給我們所須注意的相關指示與訊息。

　　我們非常重視抗體的品質，也很重視客戶的回饋與所提供的資料，所以，請儘管直接與我們分享您使用產品的經驗，若有遇到任何的問題，也請馬上告訴我們，謝謝。

　　祝好

　　泰德・阿爾科克,
　　生物合成科技

Check 好句

☑ We're sorry to hear that the antibody is not performing as well as expected.　聽到您說抗體表現不如預期，我們感到很抱歉。

☑ We expect to be able to send you the retesting results within 2 days.　希望能在兩天內給您我們的檢測報告。

電話對話

電話問貨、問匯款事 範例

人物介紹

Kitty

凱蒂，業務秘書，做事謹慎，該問的問題一個也不漏。

Roger

羅傑，客服主管，邏輯清楚，問題解決能力強，能為客戶提供又快又好的服務。

Kitty:　Hi. This is Kitty Hsu. Is Roger there, please?

Roger: Roger speaking. How are you, Kitty.

Kitty:　Just fine. I'm calling to let you know that your shipped products were found damaged on arrival.

Roger: How come? What sort of damage are you referring to?

Kitty:　Its container was damaged and product had leaked out.

Roger: I see… I think the shipment was not handled with care in transit. We'll file a claim with DHL and, and at your side, please rest assured that we'll replace the damaged product.

Kitty:　Great! When could you send out the replacement to us?

Roger: I should be able to arrange for you next Monday.

Kitty:　That would be fine. By the way, the freight should be paid by your side, right?

Roger: Correct. I'll make sure our Shipping Dept. ship with freight prepaid.

Kitty:　Thanks, Roger. I appreciate your taking care of it.

Roger: My pleasure. Is there anything else?

Kitty:　No, That's all. Thanks!

PART
1

PART
2

PART
3

PART
4

PART
5

PART
6

PART
7

電話問貨、問匯款事 範例 中文

凱蒂：嗨，我是許凱蒂，請問羅傑在嗎？

羅傑：我就是，妳好嗎，凱蒂？

凱蒂：還不錯，我打來是要跟你說，你們出的貨有到貨損壞的情況。

羅傑：怎麼會？妳說的是哪種損壞呢？

凱蒂：貨的容器破掉了，裡頭產品也漏出來了。

羅傑：瞭解…我想在運送途中貨並沒有小心處理。我們會跟 DHL 索賠，對妳這兒，請放心我們會替換損壞的產品。

凱蒂：太棒了！請問什麼時候可補出貨給我們呢？

羅傑：我應該可在下星期一安排出貨。

凱蒂：那可以，對了，運費會是由你這邊來吸收，對嗎？

羅傑：沒錯，我會確定我們的出貨部門是用預付運費來安排出貨。

凱蒂：謝謝你，羅傑，很感激你幫忙處理！

羅傑：這是我的榮幸，請問還有其他要我協助的地方嗎？

凱蒂：沒有了，就這樣，謝謝了！

電話英文短句─好說、說好、說得好！

・How come?　怎麼會？

・That would be fine.　那可以。

・My pleasure. Is there anything else?

　這是我的榮幸，請問還有其他要我協助的地方嗎？

・That's all. Thanks!　就這樣，謝謝了！

國貿知識補給站 客服與客訴

　　客服與客訴的問題，從最初始、最無傷的產品資料詢問，到廠商不認錯不賠，雙方快要撕破臉的客戶索賠情況都有，產品品質好不好、能不能用是一翻兩瞪眼的事，但其中問題處理與反應卻有不少該注意的細節，而「細節成就完美」，客訴能否完美解決，這些細節可也有頗大的影響力呢！接下來就讓我從幾個情境來瞧瞧這些細節與「眉角」：

> 「給我資料、給我答案！」

　　產品要評估、要使用，都會需要原廠的產品相關資料。在以前的年代，原廠出貨時沒附資料來就非得寫信或寫傳真去索取，但在現在這個網路世代，產品的說明書、分析報告、使用者手冊，許多廠商都已放在網路上供使用者自由下載取用，所以這告訴我們什麼呢？要索取資料、提問前，就請先在網路上找找。或許你會說「直接問比較快啊！」若是原廠人員就在你身旁，當然問會比較快，但通常這個解鈴人是在萬里之外，再加上時差因素，有辦法自己求解才是最快的！這一道功夫的用處在於找到了就立馬解決了你的問題，若沒找到，再來跟原廠反應該資料沒附、沒在網路上，或說明產品資料中並沒有你所要找的答案，請他們發來資料或回答問題，如此一來，才不會耗個一天等待，盼來的卻是原廠回說資料就在網路上或在你手上資料中的這種回答 — 而這種詢問多來個幾次，若你是代理商，也就難讓原廠肯定你這個代理商的問題處理能力了啊！

> 「產品有問題，我要換／你要賠！」

　　雖說這樣的陳述是破題，是你所要求的結論，但請注意，小紅帽的故事不會在小紅帽一出家門後，下一幕就被獵人從大野狼的肚子救出

來。一件事情的溝通與解決並不會在沒有任何交代的情況下就憑空生出。要反應事情一定要從頭說明，或許你會說「產品就是有問題，最後還不是要換！」但是你要做的功夫在於也要讓原廠有足夠的證據來判定問題出在產品、出在原廠這一方，這樣原廠才有可能答應你的換貨、賠償或退款要求。所以，貨物要是一到貨就發現短少或破損，請務必拍照留念，喔不！是拍照存證！雖說這年頭美美的照片會騙人，但貨物狀況的照片可是信服度十足呢！有佐證才不會發生像是原廠說我們有兩個人一起裝貨不會出錯，客戶也說我們有兩個人點貨不會看錯的這種狀況。若是產品不能用、檢測有問題，請也在第一時間就將發生經過、處理程序說明清楚，若是說著「我們都照著說明書上所說來操作，但結果不行用」，這樣叫不叫說明？不是，這叫有說跟沒說一樣！因為這線索完全無法對結果的判定有幫助，所以，請逐步說明，有檢測資料就附上。當你不嫌麻煩地做該做的事，提供該提供的資料，讓你所知的情況一次完整地表達出來，也才能加快原廠判定程序，縮短問題解決所需的時間。

➤ 「我前天訂單所訂的貨，就是要補先前出問題、要你們賠的貨，所以新訂貨品不應收費！」

　　請問這樣的處理有沒有問題？第一層的問題在於責任歸屬尚未釐清，所以原廠是否賠償、要否補出貨，還無法確定。第二層的問題出在處理的先後程序上，先補訂了貨，後說是該你賠的，這在原廠看來，是先斬後奏，若真急著要貨，也該在補訂當時先行說明是為了客訴案件所要補的貨。你對問題的陳述若是既完整又有理，自然說服力夠，自然容易讓對方接受你的要求。

　　客訴、抱怨的 e-mail 寫的是問題，求的是解決，反應問題時要循著道理走，不要讓指責、控訴的情緒參雜其中。e-mail 寫作時要沉得

住氣，要將問題、狀況的原委逐一寫出，將要求對方回應的點清楚說出。要怎麼樣檢測你自己的說明完不完整、有沒有道理呢？不難，請換個腦袋即可！換腦袋做不到啊？！相信我，你可以的！訣竅就是「自我檢測」！請就想像有個人拿著你的 e-mail 內容來要求你，請憑你的直覺來判斷這樣說你會不會答應？有沒有什麼你還想問的問題？若無法說服你的直覺，就請再使出你的說理能力，對原先說法再補強，補到無懈可擊的地步，你也就會有個強而有力的論述了！

PART
1

PART
2

PART
3

PART
4

PART
5

PART
6

PART
7

退貨授權書格式 範例 1

RETURN AUTHORIZATION

CONTACT INFORMATION	Return Authorization #:5568
Name: Ally Huang Company: Biochem Science, Inc. E-mail addr.: allyh@biochem.com.tw	Customer #: 0925 Sales Order #: A0815 Invoice #: 1218
SHIP TO	**Return Reason**
Biosysthesis Technology Attn.: Adele Kershner 11000 Ellsworth Rd. Ann Arbor, MI, USA	Returning unused vials which might have problems
RETURNED FREIGHT	**CSR Contact Information**
BiosynthesisTechnology's Error: YES ☐ NO ☑ If Yes, FedEx # listed below	Name: Ted Alcock Phone #: (734) 975-0000 E-mail addr. Tedalcock@BT.com

SPECIAL HANDLING/PACKING INSTRUCTIONS

Ship via Federal Express P1 International Service.
Please send the item back in the original box (or any similar insulate box), packed with frozen gels. Please ship with a minimum 30 pounds of DRY ICE PER BOX. The product must be shipped at the appropriate temperatures. Please note on CI: No Sale or transaction has occurred; value stated is for Customs Purposes Only.

Qty.	Size	Item #	Item Name	Batch #
5	200ug	1002	M1166 antibody	248

You must include a copy of this form inside the package and notate on the FedEx shipping label	Arrangement is void after 10 days.

Please ship on MONDAY or FRIDAY only.

Date: April 3, 2015	CSR Signature

退貨授權書格式 範例 1 中文

退貨授權書

聯絡資訊	退貨授權號碼：5568
姓名：黃艾利 公司名：生化科技公司 E-mail 地址：allyh@biochem.com.tw	客戶帳戶號碼：0925 銷售訂單單號：A0815 發票號碼：1218
收件人訊息	退貨原因
生物系統科技 收件人：愛黛兒‧柯許納 美國邁阿密安娜堡埃爾斯沃思路 11000 號	退回可能品質有問題的未用貨品。
退貨運費	客服關係聯絡人資訊
生物系統科技處理失誤： 是 ☐ 否 ☑ 若是，請列出 FedEx 帳戶號碼：	姓名：泰德‧阿爾科 電話號碼：(734) 975-0000 E-mail 地址：Tedalcock@BT.com

特別處理／包裝指示：

以聯邦快遞 P1 國際服務方式寄回。
請用原箱（或類似的隔熱箱）退回，配以冰凍凝膠來包裝，每箱要裝入至少 30 磅的乾冰。
此產品須在其合適溫度下運送。
請在商業發票上加註：並未有任何銷售或交易行為發生；所列金額僅供海關參考。

數量	尺寸	型號	品名	批號
5	200ug	1002	M1166 抗體	248

退貨箱中須附上此退貨授權書，FedEx 出貨標籤上亦須加註說明。	超過 10 天未退則此退貨安排即無效。
請僅於星期一或星期五退貨。	
日期：2015 年四月三日	客服關係處理人員簽名：

退貨授權書格式 範例 2

Return Authorization	Replacement required: V	RA No.: T168
1. REQUEST FROM: DISTRIBUTOR: ENVISION	Sender's Ref. No.: 1011 Name: Pamela Pan	Date: 2015-05-08 Country: TAIWAN

2. PRODUCT TYPE

☐ **Revolving door**	Article No.:_____ Serial No.:_____
☐ **Activation unit**	Article No.:_____ Serial No.:_____
☑ **Control unit**	Article No.: C81 Serial No.: AC505

3. SYMPTOM DESCRIPTION

☐ **Problem during installation**	☑ **Problem after installation**
☐ Missing parts ☐ Wrong delivery	☐ Transportation damage ☐ Other
☐ Does not work at all	☐ Does not lock / unlock
☐ Does not open	☐ Bad activation
☑ Too high / low speed	☐ Oil leakage
☑ No emergency function	☐ Action advised by manufacturer

4. DETAILED DESCRIPTION: (required)
The Controller couldn't perform stably.

Order Confirmation no.:	Installation date:

5. REPLACEMENT ORDER - Articles to be shipped

Item	Type	Article No.	Qty.
C1 Controller	Control Unit	C81	1

6. DELIVERY INFORMATION

Ship-to company/ addr.:	Envision Automation & Control 101, Sec.4, XinYi Rd. Taipei, Taiwan
Attention at site:	Pamela Chen

退貨授權書格式 範例 **2** 中文

退貨授權書	須補出貨：☑	退貨授權書號碼：T168
1. 退貨要求提出人： 經銷商：展望自動控制公司	寄件公司編號：1011 姓名：陳潘蜜拉	日期：2015-05-08 國別：TAIWAN

2. 產品類型

☐ 旋轉門	型號：＿＿＿＿ 序號：＿＿＿＿
☐ 啟動裝置	型號：＿＿＿＿ 序號：＿＿＿＿
☑ 控制裝置	型號： C81　序號： AC505

3. 狀況說明

☐ 安裝時發生問題	☑ 安裝後發生問題
☐ 零件短少　☐ 出貨有誤	☐ 運送途中受損 ☐ 其他
☐ 完全無法動作	☐ 無法鎖上／無法解鎖
☐ 無法開啟	☐ 啟動有困難
☑ 速度太快／速度太慢	☐ 漏油
☑ 緊急功能無效	☐ 原廠召回

4. 詳細說明：（必填）
控制器的反應不穩。

訂貨確認單單號：	安裝日期：

5. 替換貨訂單 — 補出貨品項

產品	類別	型號	數量
C1 控制器	控制裝置	C81	1

6. 出貨訊息

收件公司名／地址	展望自動控制公司 臺灣臺北市信義區四段 1011 號
現場收件人	陳潘蜜拉

《Dear Amy》時間

Dear Amy,

　　我當英文秘書當了好幾年了，基本上跟國貿有關的事我大多都碰過，所以這部分對我來說並不傷神，但我有一個大罩門，就是客服與客訴這一塊！對於客服與客訴這事，「對外」與「對內」我都覺得頗傷腦筋啊！「對外」的問題在於我不懂技術，下筆寫起客服問題與客訴案件時，字虛心也虛。「對內」的問題是業務、技術人員常常把要求、把要問的問題說得簡單到邏輯並不怎麼順，讓我在寫 e-mail 轉達給國外時，又一次字虛心也虛，再這樣下去，我一定會虛到得去看中醫哩！請問有沒有什麼方法能讓我在看中醫之前先補補啊？

一介弱女子 Lucy

Dear Lucy,

　　妳知道問題出在哪裡，又有求醫的心，那鐵定很快就可變成一名女硬漢呢！妳「對外」與「對內」的問題，其實同樣都是出在技術專業這部分。要補技術面的知識，請妳先分析妳可以如何做，先理出有哪些技術文件可以拿來讀，找出來後，請下功夫讀懂讀熟，並背下關鍵名詞與動詞。技術類的關鍵名詞可讓妳寫 e-mail 時，唱名唱得正確，言之有對到物，而關鍵動詞則可讓妳準確地描述出動作，不會讓原廠覺得一頭霧水。而「對內」的問題，若是妳對技術懂個六分，那麼，當業務、技術人員對妳說出不成邏輯的話時，妳會知道如何提問、如何問出漏失邏輯之處。完全不懂就無從問起，唯有懂了梗概之後，才能問出像樣的問題，才能讓雙方進行有效的溝通。此外，當業務、技術人員問妳問題，但妳反而回問他們問題時，或許他們會對妳皺皺眉頭，但請忍受一下對方眉頭皺，也趁機內部教育一下，跟對方說明當妳知道得多些，e-mail 提問就能清楚些，原廠也才更能回答得切題，更能快速地一次就替他們要到想要的答案。在妳這樣充電、疏通之後，下筆一定更穩，心一定更實，如此一來，看中醫時就請妳專心問如何補身，無須問補心了呢！

單字片語說分明

• technical [`tɛknɪk!]

adj. 技術的 relating to a particular subject, art, or craft, or its techniques

例 We are happy to provide technical support and/or training but ultimately it is the responsibility of the distributor to respond to their individual customer's needs

我們很樂意提供技術支援及／或訓練，但回應各別客戶之所需，最終仍是代理商的責任。

其他字義：專業的；技術上的；技巧上的；嚴格按照字面的

technique [tɛk`nik] n 技術；技巧；方法；手段

technically [`tɛknɪk!ɪ] adv 技術上；技巧上；嚴格來說

technician [tɛk`nɪʃən] n 技術人員（〔非正式〕techie）

常見搭配詞

a technical problem/ hitch/ glitch/ fault 技術故障

technical support 技術支援

technically speaking 嚴格來說

a technical term 術語

• instruction [ɪn`strʌkʃən]

n. 指示；命令；吩咐 a statement or explanation of something that must be done, often given by someone in authority

例 The USD payment instructions are at the bottom of the attached Proforma Invoice.

美金付款的指示說明列在所附之形式發票下方。

其他字義：指導

instruct [ɪn`strʌkt] v

instructional [ɪn`strʌkʃən!] adj 教學的；教學用的

instructive [ɪn`strʌktɪv] adj 有益的；增長知識的

instructor [ɪn`strʌktɚ] n 教練；指導者

PART 1
PART 2
PART 3
PART 4
PART 5
PART 6
PART 7

常見搭配詞

instruction manual 使用手冊　disregard/ ignore instructions 不顧指示
follow/ obey instructions 遵照指示　issue instructions 發出指示；發佈指令
disobey instructions 不依照指示　provide instructions 提供指示

・ performance [pɚˋfɔrməns]

n. 性能 the speed and effectiveness of a machine or vehicle

例 Our new version product is designed to give you unmatched performance and flexibility.

我們的新版產品是設計用來帶給您無可比擬的性能與靈活性。

其他字義：表演；表現；績效

perform [pɚˋfɔrm] Ⓥ

performer [pɚˋfɔrmɚ] Ⓝ 表演者

常見搭配詞

performance appraisal 績效評估　perform a check/ test 進行檢查
performance review 業績考評　perform a task/ duty/ service 執行任務
performance art 表現藝術　perform well/ satisfactorily 表現得好
give a performance 演出　perform poorly 表現得不好
perform an experiment 進行實驗

・ difference [ˋdɪfərəns]

n. 差別；差異 something that makes one thing or person not the same as another thing or person

例 The difference between these two products is the sensibility.

這兩個產品的敏感度不同。

其他字義：差額；分歧

differ [ˋdɪfɚ] Ⓥ

different [ˋdɪfərənt] 𝖺𝖽𝗃

differential [͵dɪfəˋrɛnʃəl] 𝖺𝖽𝗃 依差別而定的；Ⓝ（數量、價值或比率的）差別

常見搭配詞

difference of opinion 意見分歧

make a difference 產生重大的影響（尤指好的影響）

make no/ little difference 沒關係；沒有什麼影響

tell the difference 區分出來

with a difference 別具一格的

a world of difference 很大的差別

slight/ subtle/ minor difference 些微的差別

significant/ marked/ major difference 很大的差別

crucial/ essential/ fundamental difference 根本的差別

resolve/ settle your differences 解決你們之間的歧見

• complaint [kəmˋplent]

n. 投訴；抱怨 a written or spoken statement in which someone says they are not satisfied with something

例 If the Supplier determines the customer's complaint is due to mishandling on the customer's behalf, it shall be the responsibility of the Distributor to resolve such complaints.

若供應商判定此客訴的原因是客戶端操作不當，則處理此客訴的責任則歸代理商。

其他字義：疾病；不適

complain [kəmˋplen] Ⓥ

常見搭配詞

make/ file/ lodge a complaint 提出投訴

investigate a complaint 對投訴進行調查

uphold a complaint 認為投訴案有理

• compensation [ˌkɑmpənˋseʃən]

n. 賠償金 money that someone receives because something bad has happened to them

例 The company has offered a decent compensation for the missing product

這家公司對於遺失的貨，提供了還不錯的賠償金金額。

其他字義：補償；（工作的）報酬

compensate [`kɑmpənˌset] Ⅴ

compensatory [kəm`pɛnsəˌtorɪ] adj

・exchange [ɪks`tʃendʒ]

n. 交換 a situation in which one person gives another person something and receives something else of a similar type or value in return

例 We only offer an exchange of the same product in a different size or colour.

我們所提供的換貨只限於同一產品的不同尺寸或顏色。

其他字義：（貨幣的）兌換；爭吵；交流

exchange Ⅴ

常見搭配詞：名詞 exchange	常見搭配詞：動詞 exchange ＋ 名詞
exchange rate 匯率	exchange words 交談
	exchange blows/ punches 互毆
an exchange of fire/ shots 交火	exchange ideas/ views 交換意見

Part 6
索取產品相關物件篇

Unit 01 索取產品資料與樣品

單字片語一家親

新產品推出

推出；發行　launch
首次登場　debut
研討會　seminar, workshop
會議　conference
網路線上研討會　webinar
電子報　e-Newsletter
地點　venue, location
攤位　booth
參加　attend
曝光；宣傳　exposure
能見度　visibility
介紹　introduce, unveil

這些都是資料…

資料　data
文件　document
型錄　brochure
單張型錄　flyer, leaflet
分析報告　Certificate of Analysis
產品說明書　Data Sheet
物質安全資料表
Material Safety Data Sheet
技術報告　Technical Report
品管檢驗報告　QC Report
產品說明書；仿單　Package Insert

請提供樣品

要求　make a request
核准　approve
測試　test
研究　research
實驗　experiment
調查；研究　investigation
免費　for free, free of charge, at no extra charge
索價　charge for
樣品折扣　sample discount
額外折扣　additional discount

資料要用來…

研讀　perusal
評估　evaluation
參考　reference
審視　review
比較　comparison

索取的目的是要評估…

功能　function
應用　application
合適性　suitability
適合　match
標準　standard, criterion

句型

句型1 注意到新消息

It came to my attention that your company has launched sth.

例 It came to my attention that your company has launched a new campaign.

我注意到你們公司推出了一項新的活動。

句型2 請發來資料研究研究

I'd like to have sth for my perusal.

例 I'd like to have the test's finalized report for my perusal.

請給我這次測試的最終報告，我想要仔細看看。

句型3 給你連結，自行下載

Also you could download sth from the link shown below.

例 Also you could download the video from the link shown below.

你也可從下面的連結來下載影片。

句型4 產品合用喔！

The product seems to fit sth.

例 The product seems to fit my need perfectly!

這項產品看來完全符合我的需求耶！

句型5 可提供樣品給您

We'll be happy to send you the sample…

例 We'll be happy to send you the sample which is significantly different from the previous ones.

我們很樂意寄樣品給你，它跟先前的產品有很大的不同。

PART 1
PART 2
PART 3
PART 4
PART 5
PART 6
PART 7

国貿英語 溝通術
Master English | Communication for International Trade

✉ 索取產品資料 E-mail | Request for Product Data

索取產品資料 E-mail 範例 ‧要求發來產品、價格資料

Dear James,

I recently attended the International Biomedical Conference and it came to my attention that your company has launched Ace 131 Antibody for use in fluorescence microscope. I'd like to have the product's related data for my perusal. Please send me its brochure, Certificate of Analysis, Material Safety Data Sheet and any other documents that will help us know your product better.

In addition, please let me know the price of this product for my evaluation. I look forward to hearing from you.

Sincerely

Alice Liu
Creation Laboratory

單字 ShabuShabu 一小補小補！
☑ conference [`kɑnfərəns] (n.)　會議
☑ fluorescence [fluə`rɛsns] (n.)　螢光
☑ creation [krɪ`eʃən] (n.)　創造

234

索取產品資料 E-mail 範例 中文

詹姆士，您好，

　　我最近參加了國際生醫會議，在會議中我注意到您公司推出了用在螢光顯微鏡的 Ace 131 抗體，我想要仔細看看這個產品的相關資料，還請您寄給我它的型錄、分析報告、物質安全資料表，以及其他能讓我多瞭解此產品的相關文件。

　　此外，請也告訴我此產品的價格，讓我評估一下。期待收到您的回音。

謹上

劉愛麗絲
創造實驗室

Check 好句

☑ Please send me the data that will help us know your product better.　請寄資料給我，讓我多瞭解您們這項產品。

☑ In addition, please let me know the price of this product for my evaluation.　此外，請也告訴我這項產品的價格，讓我評估一下。

回覆索取產品資料 E-mail 範例 · 附上資料，也提供連結供下載

Dear Alice,

We're glad to hear that you're interested in our newly launched Ace 131 Antibody. Attached please find all your requested documents. Also you could download its all related data from the link shown below:

http://www.antibodyplatform.com/media_assets/literature/images/01-009.pdf

Moreover, you could view a replay of our New Product Launch's webinar. I trust that you'll obtain a full picture of this product.

http://www.antibodyplatform.com/?articles.view/articleNo/4101/

The list price of Ace 131 Antibody is US$ 550 and it's in stock for all sizes. We look forward to hearing from you about the order.

Kind regards,

James Jones
Antibody Platform Corp.

單字 ShabuShabu 一小補小補！
☑ link [lɪŋk] (n.)　連結
☑ moreover [mor`ovɚ] (adv.)　此外
☑ replay [ri`ple] (n.)　重播
☑ platform [`plæt͵fɔrm] (n.)　平臺

回覆索取產品資料 E-mail 範例　**中文**

PART 1
PART 2
PART 3
PART 4
PART 5
PART 6
PART 7

愛麗絲，您好，

　　我們很高興聽到您對我們新推出的 Ace 131 抗體有興趣，在此附上您所要求的所有資料，您亦可自下列連結下載所有相關的資料：

http://www.antibodyplatform.com/media_assets/literature/images/01-009.pdf

　　此外，您也可看看我們新產品發表的線上研討會重播，我相信您看了影片之後，對這一項產品就會有個完整的認識。

http://www.antibodyplatform.com/?articles.view/articleNo/4101/

　　Ace 131 抗體的定價為美金 550 元，目前所有規格皆有現貨。我們期待能收到您訂單的消息。

祝好

詹姆士・瓊斯
抗體平臺公司

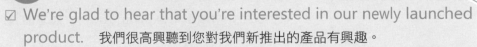

Check 好句

☑ We're glad to hear that you're interested in our newly launched product.　我們很高興聽到您對我們新推出的產品有興趣。

☑ I trust that you'll obtain a full picture of this product.　我相信您將會對這項產品有個完整的認識。

索取樣品 E-mail | Request for a Sample

索取樣品 E-mail 範例 · 請寄樣品來，通過測試後會下單

Dear James,

Thanks for the comprehensive data. I've perused all of them and watched your webinar featuring the debut of Ace 131 Antibody. The product seems to fit the research I'm conducting now. Would you please send me a sample of this product for me to test? If yes, please send it to the following address:

No. 102, JenPin Street,
SanChung District, New Taipei City,
Taiwan

If the testing is satisfying, I'll then decide the quantity I need and place an order to you. I look forward to receiving your reply.

Sincerely

Alice Liu
Creative Laboratory

> 單字 ShabuShabu 一小補小補！
> ☑ comprehensive [ˌkɑmprɪ`hɛnsɪv]
>   (adj.) 全面的
> ☑ feature [fitʃɚ] (v.) 特別介紹
> ☑ fit [fɪt] (v.) 適合
> ☑ satisfying [`sætɪsˌfaɪɪŋ] (adj.) 滿意的

索取樣品 E-mail 範例 中文

詹姆士,您好,

　　謝謝您,您所來的資料很完整,我已都研究過了,也看了您關於 Ace 131 抗體新產品發表的線上研討會重播,這一項產品看來是符合我的研究所需。請問您可以寄個樣品來讓我測試一下嗎?若可以的話,還請您寄到下面這個住址:

　　臺灣新北市
　　三重區仁品街 102 號

　　若是測試的結果令人滿意,我就會決定一下我需要的數量,再下訂單給您。期待您的回音。

謹上

劉愛麗絲
創造實驗室

Check 好句

☑ Would you please send me a sample of this product for me to test? 請問您可以寄個樣品來讓我測試一下嗎?
☑ If the testing is satisfying, I'll then decide the quantity I need and place an order to you. 若是測試的結果令人滿意,我就會決定一下我需要的數量,再下訂單給您。

回覆索取樣品 E-mail 範例・可提供樣品，但要收費

Dear Alice,

Thanks for your e-mail. <u>We'll be happy to send you the sample you requested for you to see whether it meets your research criteria</u>. However, please note that, instead of supplying for free, we offer a special discount for customers to purchase their samples of this new antibody. Please see the attached guidelines for our Sample Discount Policy.

To apply for sample discount, we need you to fill out the Sample Request Form (the 2nd attachment). Once we approve your request, we'll issue a discount code to you and please make sure to mention this code when placing your order.

If you have any additional questions, please don't hesitate to contact me. I will be happy to assist you in any way we can to fulfill your needs.

Kind regards,

James Jones
Antibody Platform Corp.

單字 ShabuShabu 一小補小補！
☑ instead of 並不；代替
☑ policy [palǝsɪ] (n.) 政策
☑ fulfill [fʊlˋfɪl] (v.) 滿足

PART
1
PART
2
PART
3
PART
4
PART
5
PART
6
PART
7

回覆索取樣品 E-mail 範例 中文

愛麗絲，您好，

　　謝謝您發來 e-mail，我們很樂意寄出您所要求的樣品給您，讓您測試看看是否合您研究所需，不過，要跟您說一聲，我們的樣品並非免費提供。對於客戶採購這一項新抗體的樣品，我們會提供一個特別的折扣，相關的說明還請您看我們附件所附上的「樣品折扣政策」。

　　若要申請這一項樣品折扣，請填寫「樣品索取表格」（請見第二個附件）。在核准您的要求之後，我們會發給您一個折扣代碼，當您下單訂購時，請務必在訂單中加註此代碼。

　　若您有任何其他的問題，請您儘管直接跟我連絡，我很樂意盡力協助您，滿足您的需求。

祝好

詹姆士・瓊斯
抗體平臺公司

Check 必 好句

☑ Please make sure to mention this code when placing your order.　當您下單訂購時，請務必在訂單中加註這一個代碼。
☑ I will be happy to assist you in any way we can to fulfill your needs.　我很樂意盡力協助您，滿足您的需求。

電話對話

電話要求樣品 範例

人物介紹

Louis

路易士，原廠業務代表，頭腦轉得快，折扣也給得快！

Diana

黛安娜，經銷商業務代表，懂得精打細算，也懂得要求！

Louis: Louis Barnard speaking. How may I help you?

Diana: Hi, Louis. This is Diana from Taiwan CoreChem Company. I noted from your e-Newsletter that you've launched a new type of collagen, right?

Louis: That's right! It was just released last week.

Diana: Many of our customers are interested in it and would like to have samples to test. Could you send us 20 bottles as free samples?

Louis: We do offer free samples but the quantity is limited to 10 per distributor.

Diana: I see... But we do need more than 10 to give to our key customers. Could you make an exception?

Louis: Hmm... I'm sure we can work something out... How about you order for the rest quantity and I exceptionally offer you a large discount like 50% off?

Diana: Well, our regular distributor discount is 40% off.... How about 60% off?

Louis: That's a large discount! But considering it's an order for samples, I could agree to offer you a one-time 60% off discount.

Diana: Great! After re-confirming the sample quantity, I'll get back to you and place our order!

PART 1
PART 2
PART 3
PART 4
PART 5
PART 6
PART 7

電話要求樣品 範例 中文

路易士：我是路易士‧巴納德，請問有什麼需要我幫忙的嗎？

黛安娜：嗨，路易士，我是臺灣科爾坎化學公司的黛安娜，我從你們的電子報上頭看到你們已經推出了一種新型態的膠原，是嗎？

路易士：沒錯！我們上星期剛推出這一個新產品。

黛安娜：我們有許多客戶對這個新產品很有興趣，想要測試一下樣品，你能寄 20 瓶免費樣品給我們嗎？

路易士：我們確實有提供免費樣品，不過每個經銷商最多只能有 10 瓶。

黛安娜：瞭解…不過我們需要超過 10 瓶的量，要給我們重要的客戶，您能破個例提供給我們嗎？

路易士：嗯…我相信我們有辦法解決這事…要不你們下單訂購多出 10 瓶的量，我破例給一個大的折扣，像是 5 折？

黛安娜：嗯，我們正常的經銷商折扣是 6 折…你看給我們 4 折的折扣如何？

路易士：這折扣很大耶！不過既然你們是訂來做樣品，我可以同意這次特別給到 4 折的折扣。

黛安娜：太棒了！等我確認我們要的樣品數量後，我就會再跟你連絡，再來下訂單給你囉！

電話英文短句—好說、說好、說得好！

· Could you make an exception?　你可以破例一次嗎？
· I'm sure we can work something out.　我相信我們有辦法解決這事。

國貿知識補給站 好「樣」的！

當你發現有家原廠的產品頗合你用、很合你所的需要，很想訂來，趕快到手，但你卻下不了下單的手？為什麼呢？因為沒用過，還真不知道合不合用，而且有些產品的價格高，要是訂進來不能用，那不就虧大了？這時候就該動動手指頭，寫寫 e-mail 或是在原廠網站上直接申請樣品了。原廠對於客戶索取樣品，會有幾種不同的處理與要求方式，我們就從最乾脆的一切費用全免，到也是乾脆的一切費用照付的情況來一一看下去囉！

➢ 不用不用，送你用！

這是索取樣品的最大好康，原廠願意提供小包裝樣品，讓客戶測試、試用，這種好康唯一的成本就是運費了。一般說來，不管原廠是免費提供某一數量的型錄、行銷小贈品，或是樣品給經銷商或客戶時，運費一律都是收件人自行負擔。例如："We're willing to send you a sample for free of charge, but you need to bear the shipping costs."（我們願意免費提供樣品給您，但您得自付運費。）

➢ 不用錢，但要回報！

為了要為替其產品打開市場、增加銷量，也為了擴增產品在不同實驗、儀器或專案的使用領域，並蒐集結果報告，有些原廠會願意提供免費樣品，但要求回饋檢測、使用報告。例如："We would like you to please fill out our free sample Feedback Form. This allows us to know your thoughts as well as your experience with our products."（請您填寫免費樣品的「回饋表」，讓我們知道您對我們產品的想法與使用的經驗。）

➤ 要錢，但可打個折！

　　在產品單價不低、成本掌控、樣品供量控制的考量下，有些原廠並不提供免費樣品，而是要求客戶下單購買所要的樣品規格，但在價格上有優惠，可提供樣品折扣。當然，所謂的樣品，就是先前沒用過，第一次用來打樣試試，所以這種折扣也就只能讓每個客戶或最終使用者使用一次。那原廠要如何監控是否有濫用樣品名義下單的情況呢？冤有頭、債有主，誰用了樣品當然也要報上名來，所以原廠會要求客戶在索取樣品時，也要將身家所屬公司、機構、聯絡資料一併送上，供其審核。下兩頁裡頭的樣品政策範例，說的就是這個情況。

➤ 哪有不要錢的！

　　樣品？沒有！要就請直接下單！原廠的考量有成本、產品價值與屬性，或是原廠有提供了配套的問題解決與退貨方案，例如："Please note that we do not provide free samples of any of our products, because we have a full and generous returns policy."（因我們有完整且慷慨的退貨政策，所以我們所有的產品都不提供免費樣品。）

　　最後，要來說說原廠的期待與好客戶的責任與義務。樣品來了，用了…然後呢？請記得要將結果回饋給原廠。或許你覺得使用得不順、不合，不會再買了，為什麼要回？其實，你多一道工，寫個e-mail 將狀況回給原廠，原廠就有可能以其專業跟你說明不合用的原因，也或許可告訴你什麼樣的產品與規格適合你。所以，你給一個回饋，可能讓你自己得到些新的訊息與機會，也讓原廠多了個學習與改善的機會，對雙方都好，何樂而不為？再者，這也是紮馬步的功夫！將事情做得完整，有頭有尾，這是對人對事負責的態度養成訓練。索取樣品這事不大不難，若能要求自己這類事都做足做好，等大事來

了，以你已經練就的工作態度來處理，用心且盡力，那大事就更有機會帶來大成就了！

樣品折扣政策 範例

Antibody Platform
Sample Discount Policy

The purposes of providing samples are for customers to try our newly launched products and, especially, let our main competitors' customers convertto buy from us.

Rules:
➢ Samples are offered 60% off discount for the small size of antibody.
➢ Customer information is required:
  ■ Name
  ■ Institution
  ■ E-mail address
  ■ Phone number
➢ Only 3 samples are available for per end user to request. If the end user is a high-volume user, extra samples will be approved case by case.
➢ Sample Request Restrictions
  ■ No A1 seriesproducts
➢ If only the large size is available, sample requests will be reviewed case by case.

PART
1

PART
2

PART
3

PART
4

PART
5

PART
6

PART
7

樣品折扣政策 範例 中文

抗體平臺公司

樣品折扣政策

樣品提供的目是為了讓客戶試用我們新推出的產品，特別是讓主要競爭廠牌的客戶能轉而向我們購買。

規則：

➤ 抗體小包裝的樣品可享 4 折的折扣。

➤ 須提供客戶資料：

■ 姓名

■ 機構名

■ E-mail 地址

■ 電話號碼

➤ 每位最終使用者最多只能申請三種樣品。若此最終使用者所用的量大，需要更多樣品，則將以個案處理。

➤ 索取樣品的限制

■ A1 系列產品不提供樣品

➤ 若產品只有人包裝可供，則樣品索取將以個案處理。

樣品索取程序與表格 範例

Procedures for Requesting a Sample

■ Please fill out the Sample Request Form.

■ Please send in the Form to samplerequest@antibodyplatform.com

■ Your e-mail will be received by our Sales Representatives.

■ All requests will be reviewed. Please keep in mind that some samples may not be approved due to production cost.

■ Once approved, a quote no. for the requested sample will be issued.

■ While placing an order with the sample, please include the quote no.

Sample Request Form

Name of Distributor:		Date of Request:	
Distributor Contact:		Contact E-mail:	
Date of Follow-up:		By Whom:	
Name of End User:		Date of Request:	
Institution:	(Name of Lab, Department/Division)		
Institution Addr.:			
Purpose of Sample(s):			
List of Sample Needed (Max. 3):			

索取樣品的程序與表格 範例　中文

樣品索取程序

■ 請填寫「樣品索取表格」。

■ 請將此表格發送至 samplerequest@antibodyplatform.com。

■ 您的 e-mail 將由我們的業務代表處理。

■ 我們會審核所有的索取要求。請注意，若干樣品可能因生產成本考量而無法核准提供。

■ 索取案件一經核准，我們將發給您該樣品的報價單號。

■ 請在您下單訂購樣品時，列出此一報價單號。

樣品索取表格			
經銷商名：		要求日：	
經銷商連絡人姓名：		連絡人 E-mail：	
追蹤日期：		追蹤人員：	
最終使用者姓名：		要求日：	
機構名：	（實驗室及所屬部門名稱）		
機構地址：			
索取樣品之目的：			
列出所需的樣品（最多三項）：			

《Dear Amy》時間

Dear Amy，

　　請聽我說件小事…上個星期我們公司有個業務要我跟國外原廠要產品資料，我問他是什麼樣的資料，他回我「就是產品資料啊！」，在得不到更多訊息的情況下，我就這麼寫 e-mail 給國外了，果然，原廠也問回我這樣的問題，業務聽我說之後，先是罵了一下原廠，才再跟我說要的是操作手冊，所以我再寫給原廠，但原廠又回我不太清楚我要的是什麼…看到原廠這樣回，我肩頸的肌肉就僵硬起來了，因為我知道又要再去聽業務罵一次，又要讓我心情不好了！這種小事實在常發生，我不能控制業務不罵人，那我怎樣才能控制我可以有好心情呢？

外面有太陽，可是我心中在下雨的 Michelle

Dear Michelle,

　　首先要來稱讚妳一下，妳想求解的方向是對的！我們確實不能控制別人，但絕對有辦法控制自己的心情！產品資料有好幾種，雖然看了名稱大致就能知道內容類別，但有些原廠的文件名稱確實跟其他廠家有點不同，所以，當妳或業務不清楚原廠該文件確實的名稱時，那就請妳多說說狀況，看是收到貨了，該附的產品說明書沒附（這樣原廠就知道要給你本來應要隨貨的資料），還是會列批號的的貨品檢測報告沒附（這樣原廠就不會發給妳一般的、不分批次的產品說明書），或者是要哪一種操作步驟的資料（這樣原廠就知道要給妳哪種詳細步驟與參數的操作資料）。請記得，妳多提供些線索，就更能確保原廠會發來對的資料。再來讓我們看看要怎麼解決掉這個業務！不是啦！是說業務若有這樣的態度，我們要怎麼應對？大原則，從他的利益出發來溝通！當妳要業務提供線索時，請說明先前就有過原廠來來回回問到底要什麼樣資料的經驗，所以，若要快快拿到資料，請業務多提供訊息。在妳盡到責任，做到該提醒、該做的事之後，就請笑看結果一不是真笑喔！是指坦然以對，當妳心上坦然，妳就能夠較不受外來負面情緒影響，那就可以控制自己有個好心情了喔！

單字片語說分明

· launch [lɔntʃ]

v. 推出;發行 to start selling a new product or service to the public

= release, put out, come out

例 We are very excited to announce that we will be launching a brand new website next week.

我們很興奮地要來宣佈一件事,那就是我們將在下星期推出全新的網頁。

其他字義:發射;發動;發起;展開

launch [lɔntʃ] n.　　　　　launcher [`lɔntʃɚ] n. 發射裝置;發射器

常見搭配詞:launch 名詞

lanch date 發行日　　　　launch event 產品發表會

· debut [dɪ`bju]

n. 首次登場 the first time that a performer or sports player appears in public

例 Our new interactive product will be making its debut on Sunday!

我們新的互動產品將要在星期天首次登場!

debut [dɪ`bju] adj.

常見搭配詞

make sb's debut	acting/film debut	singing debut
首次登台	首次演出	首次演唱
directorial debut	debut album	debut novel
首次執導	首張專輯	首部小說

· seminar [`sɛmə͵nɑr]

n. 研討會 a meeting at which a group of people discuss a subject

= workshop

例 We cordially invites you to attend our seminar which is designed to highlight strategies, techniques, services and new products.

我們誠摯地邀請您參加我們所舉辦的研討會，此研討會特別針對策略、技術、服務與新產品這幾個主題而設計。

其他字義：（大學）專題討論課

常見搭配詞

a joint seminar 聯合研討會　　　an international seminar 國際研討會

· **webinar** [wɛmə͵nɑr]

n. 線上研討會 a talk on a subject which is given over the internet, allowing a group of people in different places to watch, listen and sometimes respond on the same occasion

= web + seminar

例 You are officially invited to attend our webinars, covering the topics about the most advanced techniques for next-generation products.

我們正式地邀請您參加我們的線上研討會，主題涵蓋了下一代產品最先進的技術。

常見搭配詞

a live webinar
直播線上研討會

a post-webinar online survey
線上研討會的會後線上調查

· **attend** [ə`tɛnd]

v. 參加 to be present at an event or activity

= workshop

例 You are receiving this email because you placed an order, signed up online, or attended a tradeshow.

您會收到這份 e-mail 是因為您有跟我們訂過貨、在線上登錄過，或是曾經參加過貿易展。

其他字義：照料；看護；（作為結果）伴隨

attendance [ə`tɛndəns] ⓝ 出席;出席人數;照料

attendee [ə`tɛndi] ⓝ 出席者

attendant [ə`tɛndənt] ⓝ 服務員;侍從;看護員。ⓐⓓⓙ 隨之產生的;照料的

常見搭配詞

attend to 處理;服務

in attendance (正式)出席(重要或官方的活動)

· exposure [ɪk`spoʒɚ]

n. 曝光;宣傳 things that are written or said about a person, product, event etc. that make them well known

例 Our company have got a lot of brand exposure through attending exhibitions.

我們公司透過參展取得了很高的品牌曝光度。

其他字義:暴露;揭露;接觸;朝向;底片;曝光時間

expose [ɪk`spoz] ⓥ

exposition [͵ɛkspə`zɪʃən] ⓝ 說明;闡述;博覽會

常見搭配詞

media exposure	press exposure	television exposure
媒體曝光	新聞曝光	電視曝光

· visibility [͵vɪzə`bɪlətɪ]

n. 能見度 the distance that you can see, depending on conditions such as the weather or the place that you are in

例 We are a web design company using a Low Cost- High Quality Model to increase your website visibility.

我們是一間網路設計公司,利用一種低成本—高品質的模型來增加你們的網路能見度。

visible [`vɪzəbl̩] ⓐⓓⓙ 看得見的;明顯的

visibly [ˋvɪzəblɪ] adj 明顯地

常見搭配詞

public visibility	international visibility	poor/ low visibility
大眾能見度	國際能見度	能見度低
good/ high visibility	clearly visible	a visible sign
能見度高	清晰可見	明顯跡象

- approve [əˋpruv]

v. 核准 to give official agreement or permission to something

例 If you need a certain discount to be able to compete, please let me know the details and I will try to get it approved.

若是要給到某個折扣才能競爭得了，那就請你將細節告訴我，我會盡量爭取核准該折扣。

其他字義：贊同；讚許

approval [əˋpruvl] n
approving adj 贊同的；讚許的

approved [əˋpruvd] adj 被認可的

常見搭配詞

meet with sb's approval	a nod approval
得到某人的認可或讚許	點頭讚許
a smile approval	a murmur of approval
微笑讚許	一陣輕輕的讚許聲
on approval	subject to sb's approval
不滿意可退貨	須得到某人的批准

- comparison [kəmˋpærəsn]

n. 比較 the process of considering how things or people are similar and how they are different

例 We've tried hard to collect related information so as to make comparisons between similar products in the market.

我們很努力地收集相關資料，這樣才能將市場上相似的產品拿來比較。

compare [kəm`pɛr] Ⓥ

comparable [`kɑmpərəb!] adj 可比較的；差不多的；比得上的

comparative [kəm`pærətɪv] adj 比較的；相比的；相比較而言的；相對的

常見搭配詞

make/ draw a comparison	by comparison (with)
做比較	（與…）比較起來
in comparison	bear/ stand comparison (with)
相較之下	比得上
there's no comparison	compare and contrast
不能相提並論	比較異同

Unit 02

給個證明

單字片語一家親

驗明正身！

自由銷售證明
Certificate of Free Sale
製售證明
Certificate of Manufacture and Free Sale
授權書 / 代理證
Power of Attorney
Letter of Authorization
Authorization Letter
Certificate of Authorization
產地證明　Certificate of Origin
登記證
Certificate of Registration
營利事業登記證
Business Registration Certificate
進口許可證　Import Permit
出口許可證　Export Permit
健康證明　Health Certificate
衛生證明　Sanitary Certificate
CE 標記　CE Marking
ISO 認證　ISO Certification
聲明函　statement

審核

主管機關　authority
（政府機構的）部　Ministry
（政府機構的）局　Bureau
駐外代表處
Representative Office
商會　Chamber of Commerce
證明　certification
公證　notarization
認證　authentication
驗證　legalization
授權　authorization
給予權力　entitle, grant

有限制條件的產品

管制產品　regulated products, restricted products
專利保護產品
patent-protected products
註冊產品　registered products
GMP 認證產品
GMP certified products

PART
1

PART
2

PART
3

PART
4

PART
5

PART
6

PART
7

句型

句型 1 請給個權利

Taking this opportunity, we'd like to request that you agree to grant us sth.

例 Taking this opportunity, we'd like to request that you agree to grant us all necessary rights.

藉此機會，我們要來請求您同意授予所有必要的權利給我們。

句型 2 任命、派任

We will be delighted to appoint you as our...

例 We will be delighted to appoint you as our contractor.

我們很高興能委任您們公司為我們的承包商。

句型 3 附上草約

A copy of the draft agreement is attached...

例 A copy of the draft agreement is attached with relevant sections highlighted by underlining.

附上草擬的合約，裡頭提到的相關條款都有以底線標示出來。

句型 4 官方要正式文件

Our authority requires us to submit a statement certifying...

例 Our authority requires us to submit a statement certifying that we're your exclusive distributor.

我們的主管機關要求我們提出聲明函，證明我們是您們的獨家代理商。

句型 5 這些文件都要！

The following documents are required to be submitted...

例 The following documents are required to be submitted within 10 days.

下列文件必須在十天內提交。

請求授權證明 E-mail | Request for Certificate of Authorization

請求授權證明 E-mail 範例 · 請求獨家代理

Dear Adam,

   Our customer will undertake an annual project which contract will be worth US$ 20,000 or so. We're planning to promote and supply your product to the customer. In order to be a supplier, we are required to pass their audit and submit manufacturer's Letter of Authorization. We've placed many orders to you in the past year. Taking this opportunity, we'd like to request that you agree to grant us exclusive distributorship.

   For your information, with our efforts focused in biomedical fields since 1990, we have built a large customer base which includes the majority of Taiwan's leading hospitals, universities, research institutes, and pharmaceutical manufacturers. Also we have earned a good reputation in the industry in providing professional and quality service for our customers.

   Please let us have your comments. Hope to hear positive news from you.

Kind regards,
Betty Hsu
Goal Biomedical, Inc.

單字 ShabuShabu 一小補小補！
☑ audit [ˋɔdɪt] (n.)　審核
☑ exclusive [ɪkˋsklusɪv] (adj.)　獨家的

請求授權證明 E-mail 範例 中文

亞當，您好，

　　我們有一個客戶準備要執行年度專案，合約金額在美金 20,000 元左右，我們打算推薦您的產品給客戶。若我們要成為此案的供應廠商，則須通過客戶的稽核，提出原廠的授權證明。在過去這一年裡，我們已跟您下了好多筆訂單，藉此機會，我們要來請求您同意授予獨家經銷權給我們。

　　在此跟您說明一下，我們公司自 1990 年即投注努力在生醫這個領域，我們已建立了廣大的客戶基礎，在臺灣居領先地位的各大醫院、大學、研究機構及藥廠，大多都是我們的客戶。因為我們提供給客戶既專業又優質的服務，所以在業界也享有好聲譽。

　　請告知您的意見，希望能聽到您來的好消息。

祝好

許貝蒂
高爾生醫公司

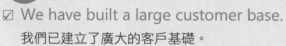

Check 好句

☑ We have built a large customer base.
　我們已建立了廣大的客戶基礎。

☑ Also we have earned a good reputation in the industry.
　我們在業界也享有好的聲譽。

回覆請求授權證明 E-mail 範例・同意給予獨家代理、先談合約

Dear Betty,

　　We will be delighted to appoint you as our exclusive distributor in Taiwan. We're impressed indeed with your customer coverage and also your sales performance in the past year.

　　Before we can issue a Letter of Authorization, we will need to sign a Distribution Agreement with you. This will be an exclusive distributorship. We would like to offer you a 20% distributor discount with this to be re-evaluated at the end of 2015. Eden Technology and Goal Biomedical also need to agree minimum revenues for the next 3 years.

　　A copy of the draft agreement is attached for you to read. If you have any questions regarding the draft please let me know. Once the agreement has been signed, then we can issue the Letter of Authorization to you.

Best regards,
Adam Springer
Eden Technology

單字 ShabuShabu 一小補小補！
☑ impress [ɪmˋprɛs] (v.)　給…極深的印象
☑ coverage [ˋkʌvərɪdʒ] (n.)　覆蓋範圍
☑ draft [dræft] (n.)　草稿

回覆請求授權證明 E-mail 範例 中文

貝蒂，您好，

　　我們很高興能委任您們公司成為我們在臺的獨家代理：您們這一年來業務的客戶數與業績表現，著實讓我們印象深刻。

　　在出具授權證明之前，我們必須先跟您們簽訂經銷合約，合約中會給您們獨家代理的條件，並給予 8 折的折扣，2015 年底時會重新評估折扣率。伊甸科技與高爾生醫亦須談定接下來三年的最低業績金額。

　　在此附上草擬的合約供您詳閱，若是您對此草約有任何問題，請告訴我們。待此合約簽訂後，我們就會出具授權證明給您。

祝好

亞當・斯普林格
伊甸科技

Check 好句

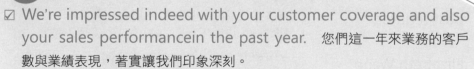

☑ We're impressed indeed with your customer coverage and also your sales performancein the past year.　您們這一年來業務的客戶數與業績表現，著實讓我們印象深刻。

☑ We also need to agree minimum revenues for the next 3 years.　我們亦須談定接下來三年的最低業績金額。

請求進出口許可證 E-mail | Request for Import & Export Permits

請求申請進口許可證所需資料 E-mail 範例 **1** · 聲明函

Dear Eason,

Thanks for your quotation. We would like to place an order for your following product: Catalog No.: SERAB101 / Product: PNH Serum, 100ml

However, this product is a regulated item and we need to apply for its import permit. <u>Our authority requires us to submit a statement certifying the following points:</u>

1. The product is not a restricted item in your country and can be exported.
2. The product is not infected by Hepatitis B, Hepatitis C, HIV1 & HIV2.

Please confirm whether you could issue such statement to us. We wait for your reply. Thanks.

Best regards,

Elizabeth Wang
BioNote Group

單字 ShabuShabu 一小補小補！
☑ regulate [`rɛgjə‚let] (v.)　管理；控制
☑ restricted [rɪ`strɪktɪd] (adj.)　受限制的
☑ infected [ɪn`fɛktɪd] (adj.)　受感染的

請求申請進口許可證所需資料 E-mail 範例 **1** 中文

伊森，您好，

　　謝謝您發來詢價，我們想要跟您訂購下列產品：
型號：SERAB101／產品：PNH 血清，100ml

　　然而，此產品為管制品項，因此我們須申請進口許可證。<u>我們的主管機關要求提交一份聲明書，以證實下列各點：</u>

1. 此產品在您們的國家非屬管制產品，可出口。
2. 此產品未受 B 型肝炎、C 型肝炎、HIV1 及 HIV2 感染。

　　還請您確認是否可出具此聲明書給我們。等候您的消息。

祝好

王伊莉莎白
拜爾諾集團

 好句

☑ This product is a regulated item and we need to apply for its import permit.　這項產品為管制品項，因此我們須申請進口許可證。
☑ Please confirm whether you could issue such statement to us. 還請您確認是否可出具此聲明書給我們。

Dear Elizabeth,

Thanks for your inquiry. We can supply this material to you. Its price for it would be US$145/10ml. We look forward to your order. Please note that, since it is a regulated item, the following documents will be needed in order for us to proceed with an order of it and apply our export permit:

1. A notarized copy of Import Permit: The copy must be translated into English if supplied in a foreign language.
2. A Certification: A certificate states that the substance consigned to the authorized importer is to be applied exclusively to medical or scientific use within the country of destination and it will not be re-exported from such country,

Please note there is an additional US$150 document fee for regulated products on top of the cost of the product and the shipping fees. If you have any questions, please do not hesitate to contact me.

Sincerely,

Michael Neal
Rock Immunochemicals, Inc.

單字 ShabuShabu 一小補小補！
☑ proceed [prə`sid] (v.)　進行
☑ substance [`sʌbstəns] (n.)　物質
☑ destination [ˌdɛstə`neʃən] (n.)　目的地

請求申請進口許可證所需資料 E-mail 範例 **2** 中文

伊莉莎白，您好，

謝謝您來信詢價，我們可供貨給您，產品的價格為 US$145/10ml，期待能收到您的訂單。要請您注意一下，因為這是管制產品，所以，在我們處理訂單前，我們還需要下列的文件：

1. 經認證的進口許可證影本：若許可證為外國語言版本，則此影本須翻譯為英文。

2. 證明：須提出一份證明，說明出給授權進口商的產品，僅能在目的地國家境內在該國境內供醫學或科學用途使用，不會轉出口至其他國家。

請注意對於此類管制產品，除了產品成本與運費之外，須另加 US$150 的文件費。若您有任何問題，請儘管與我們連絡。

謹上

麥克·尼爾
磐石免疫化學公司

Check 好句

☑ Since it is a regulated item, the following documents will be needed.　因為這是管制產品，所以我們需要下列文件。

☑ There is an additional fee on top of the cost of the product.　除了產品成本之外，須加另一筆額外費用。

申請產品登記 E-mail | Apply for Product Registration

請求產品登記所需資料 E-mail 範例 · 自由銷售證明、委託書

Dear Ruby,

　　We are preparing to apply the registration for your products in Taiwan. <u>The following documents are required to be submitted on application</u>:

　　1. Certificate of Free Sale
- The Certificate should be issued by the appropriate competent national authority, giving assurances that your products can be sold freely.
- It should be notarized by our notary public in the US.
- It should have a validity of 2 years.
2. Power of Attorney (POA)
- The Power of Attorney could be issued by the manufacturer.
- It should have a validity of 1 year.

　　Please prepare these documents and let us know how long it will take for you to send them to us. Thanks.

Best regards,

Jade Hsin
Benchmark Technology

單字 ShabuShabu 一小補小補！
- ☑ appropriate [ə`proprɪˌet]
  (adj.) 適當的
- ☑ competent [`kɑmpətənt]
  (adj.) 有法定資格的
- ☑ notary public 公證人

請求產品登記所需資料 **E-mail** 範例 中文

露比，您好，

　　我們目前正準備著要替您們的產品在臺灣辦理登記，<u>申請時即須提交下列文件</u>：

　　1. 自由銷售證明
　　- 此證明須由您們所屬國家主管機關出具，確保您們的產品可自由銷售。
　　- 此證明須由我們在美國的公證機構認證。
　　- 此證明須有 2 年效期。
　　2. 委託書
　　- 委託書由製造廠出具即可。
　　- 委託書須有 1 年的效期。

　　請準備這些文件，也請告知何時可寄給我們，謝謝。

祝好
邢潔德
標竿科技

Check 好句

☑ We are preparing to apply the registration for your products in Taiwan.　我們目前正準備著要替您們的產品在臺灣辦理登記。

☑ Please prepare these documents and let us know how long it will take for you to send them to us.　請準備這些文件，也請告知何時可寄給我們。

回覆請求產品登記所需資料 **E-mail** 範例 · 哪些產品、何人認證

Hi Jade,

Our Regulatory Affairs Department is putting together the information you require. Could you confirm which products are to be listed on the Free Sale Certificate? For your information, to obtain a Certificate of Free Sale requires a $75 document fee and takes approximately 2 weeks.

Additionally, I assume your request for the certificate to be notarized will require notarization by the Taiwanese chamber of commerce in the US. Please confirm.

For the Power of Attorney, we can provide a more recent letter of authorization. Let us know if this will suffice. Thanks.

Best regards,

Ruby Lopes
Milestone Science Corp.

單字 ShabuShabu 一小補小補！
☑ affair [əˋfɛr]] (n.)　事務
☑ assume [əˋsjum] (v.)　以為；認為
☑ chamber of commerce　商會

回覆請求產品登記所需資料 E-mail 範例 中文

潔 ，您好，

　　我們的法務部門正在彙整您所需的資訊，能否請您確認要列在自由銷售證明中的產品有哪幾項呢？在此讓您知道一下，申請自由銷售證明須收取美金 75 元的文件費，約需兩個星期的時間。

　　此外，關於您對自由銷售證明認證的要求，我想您指的是要由在美國的臺灣商會來辦理認證，還請您確認這一點。

　　至於委託書一事，我們可以更新先前所給您的授權書，請告知這樣是否可符合您所需，謝謝。

祝好

露比・洛普斯
里程碑科學公司

必 Check 好句

☑ To obtain a Certificate of Free Sale requires a $75 document fee and takes approximately 2 weeks.　申請自由銷售證明須收取美金 75 元的文件費，約需兩個星期的時間。

☑ Let us know if this will suffice.　請告知這樣是否足夠。

Goal Biomedical, Inc.

33F, No. 310, Ruiguang Road, Neihu District, Taipei City 114, Taiwan

Tel: 886-2-8750-0001

Fax: 886-2-8750-0002

Date: April1, 2015

TO WHO IT MAY CONCERN

By this letter, we state that Goal Biomedical, Inc. with address at: 33F, No. 333, Ruiguang Road, Neihu District, Taipei City 114, Taiwan, is an authorized representative of Eden Technology in the Territory of Taiwan, to sell and promote all products made by our company. The offices of Eden Technology are currently located in 1550 S. Milpitas Blvd. Milpitas CA 95035 making and commercializing products under the registered trade names Eden Technology.

This authorization is valid until December 31, 2015.

Adam Springer

International Sales and Distributor Manager

Eden Technology

授權證明格式 範例 中文

高爾生醫公司
114 臺灣臺北市內湖區瑞光路 310 號 33 樓
電話：886-2-8750-0001
傳真：886-2-8750-0002

日期：2015 年四月一日

敬啟者

茲以此信函聲明高爾生醫公司，設址於 114 臺灣臺北市內湖區瑞光路 310 號 33 樓，為伊甸科技在臺灣地區所授權的代理商，負責銷售與推廣我們公司所生產的產品。伊甸科技設址於 95035 美國加州密爾必達市密爾必達大道 1550 號，生產與銷售以伊甸科技為註冊商標名之產品。

此授權至 2015 年十二月三十一日前有效。

亞當 · 斯普林格
國際銷售與經銷商部門經理
伊甸科技

製售證明格式 範例

NEW YORK STATE
DEPARTEMNET OF HEALTH

CERTIFICATE OF FREE SALE/MANUFACTURE

This is to certificate, based upon information and beliefs, that the following product manufactured for Milestone Science Corp., whose principal office is located at 1001 168th Street, Flushing, NY 11358, is freely sold without restrictions throughout the United States of America and is otherwise manufactured, according to Good Manufacturing Practices (GMP) in compliance with all applicable international, federal, state, or local laws, rules, and regulations relating to the manufacturer, marketing, and sale of such product:

Product	Item	Size
25-Hydroxy Vitamin D	MMS-101	96 wells

NEW YORK STATE DEPARTEMNET OF HEALTH

製售證明格式 範例　中文

紐 約 州 衛 生 局

製 售 證 明

基於所提呈資訊與信任，茲以此證書證明由里程碑科學公司，設址於 11358 紐約法拉盛區第 168 街 1001 號，所製造之下列產品，可不受限制地在美國全區自由銷售，其生產符合優良製造規範，合乎與此產品製造、行銷與銷售相關之所有國際、聯邦、州立與地方法律、規則與條例：

產品　　　　　　　　　　型號　　　　　　規格

25一羥基維生素 D　　　MMS-101　　　96 孔盤

紐約州衛生局

國貿英語 溝通術
Master English Communication for International Trade

📞 電話對話

電話討論授權書細節 範例

人物介紹

Toni

唐妮，經銷商業務秘書，頭腦清楚，談事情條理分明！

Albert

亞伯特，原廠業務助理，做事偶有疏漏，不過，改正動作還算快！

Toni: Hi. Is Albert in?

Albert: Yes, this is Albert. Is this Toni Lin?

Toni: It sure is! I'm calling about the Letter of Authorization you sent me yesterday. I'd like to discuss some points with you.

Albert: Sure. Is there anything missing to be mentioned in it?

Toni: It is. First of all, the Letter of Authorization should be printed on your company letterhead.

Albert: No problem! I can do that.

Toni: Great! Next, there's no signature on it.

Albert: Oh... It's my oversight. Sorry! I'll have it signed by our International Marketing Manager.

Toni: Great! There's still one last thing. The Letter's validity is one year only. We hope that it can be extended for one more year, which will enable us to make more thorough marketing strategies.

Albert: I'll have to get back to you on that. I'm not in a position to authorize that. I'll talk to my supervisor.

Toni: That's fine. Once you have an answer, please let me know.

Albert: Certainly!

電話討論授權書細節 範例 中文

唐　妮：嗨，請問是亞伯特嗎？

亞伯特：是的，我就是，妳是林唐妮嗎？

唐　妮：沒錯！我打來是要談談我昨天發給你的授權書，有幾點我想要跟你討論一下。

亞伯特：好的，我有漏了什麼嗎？

唐　妮：有的，首先，這份授權書應該要列在有你們公司抬頭的信紙上。

亞伯特：沒問題，這我可以做。

唐　妮：謝謝！接下來，它上頭也沒有簽名耶。

亞伯特：喔…是我漏了，不好意思！我會請我們的國際行銷經理簽名。

唐　妮：太好了！還有最後一點，這份授權書的效期只有一年，我們希望可多一年，這樣我們才好擬訂更完整的行銷策略。

亞伯特：這一點我得再回妳，這部分我無權決定，我會去請示我的主管。

唐　妮：好的，請你一有消息就告訴我。

亞伯特：一定！

電話英文短句一好說、說好、說得好！

· It sure is!　沒錯！
· I'd like to discuss some points with you.　有幾點我想要跟你討論一下。
· I'll have to get back to you on that.　這一點我得再回你。
· Certainly!　一定！

國貿知識補給站　擺平申請正式文件事

　　碰到索取官方出具文件或辦理登記這類正經事，你想偶爾不正經一下是鐵定行不通的！要想快快擺平它，得要你完完全全清楚你所要文件的所有相關要求，若是漏了一點，然後就這樣送件，光那一點就會整個翻盤，整件事從頭再來一次，除了消耗國外與你的精力，也浪費了時間與申請費用成本，更重要的是可能延誤了交貨或是業務推廣的時機。究竟碰到這類正式文件的申請時，我們該注意哪些地方呢？讓我們一起來修一修性子，培養一下嚴謹行事的處事態度吧！

1. 背景說明

　　要做一件事之前，一定要讓對方與你有相等的資訊量，你知道四個要點，就不要只跟對方說個一點…很多時候在你怪罪國外辦事不夠力、不夠快時，其實是你的要求並沒有說得很清楚，沒能讓國外原廠確實知道為什麼要提供文件，何時一定得提供，不提供的後果會如何…將背景狀況說明清楚後，國外原廠也才有可能配合，以你認為該有的重視程度，以該有的處理、跟催速度來辦理。

2. 要的文件有哪些？

　　與人相處，說話時若是只選一部分的實情來說，不要懷疑，那就算是說謊，而在與國外原廠溝通所需文件有哪些時，若你只說個部分，那倒不是說謊，但絕對不是負責任的表現。所以，請務必先找出、理出我們主管機關要求的文件到底有哪些，一次表列、條列給原廠，切勿這星期要這項證明，等原廠發來證明後，才又告訴原廠還缺另一份聲明書，待送件辦理申請後，再跟原廠說還有一份檢驗文件也需要…若是讓這樣的情況發生，除了辦理申請的時間會拉得很長之外，原廠對你、對你所代表公司的辦事能力，也難免要大打折扣了！

3. 文件由誰出具？由誰認證？

　　整理出所需文件後，請逐一釐清哪些需國外官方出具，哪些是原廠自行出具即可。若要官方文件，是要由哪個主管機關出具？若是有文件須認證，該由哪個單位認證？這些都是在要求國外原廠時，必須先理清楚的要點。

4. 文件的內容

　　所出具的證明或聲明，若我們的主管機關有要求一定要列出什麼樣的句子與敘述，請務必在要求國外原廠時一併提出，若是可跟我們的主管機關要到樣本，那就更好了，就可直接提供給原廠，讓原廠比照辦理，避掉漏失任何訊息的風險。

　　不管是提出要求的一方，還是被要求的一方，在辦理這類正式文件申請時，請恪盡通報進度之責，因為茲事體大又耗時，時間一拉長，不管是提出要求的當事人，或是負責執行的公司內部或官方人員，都很有可能忘了處理、忘了送件、忘了催促，所以除了申請程序本身就耗時之外，再加上人為的選擇性遺忘，很有可能幾個月都過去了卻始終沒有門鈴響，這樣的延宕雖不至於讓你人老珠黃，但卻很有可能讓你碰到在最後期限前卻突然發現快來不及了的緊急狀況。所以，在申請正式文件這事上，若是要讓我們自己不要在最後關頭才瘋狂索命連環追，而是要能長保優雅地就這麼催一下、盯一下，究竟要怎麼做呢？最高指導原則就是：請記得三不五時就去催一下、盯一下吧！

國貿英語 溝通術
Master English Communication for International Trade

《Dear Amy》時間

Dear Amy，

　　我從上一個公司離職的最大原因，就是工作太重、太忙，尤其是我還要負責跟國外原廠溝通辦理醫療儀器登記的事，這件事辦來沒完沒了，三天兩頭就要寫個長長的 e-mail 來跟原廠說明我們的要求，而且辦完了一項儀器，下一個儀器又要開始申請。我自認工作很認真，但這類工作真的讓我有些吃不消，所以我捨下同事情，決定離職。到了這家醫藥公司，倒還能準時下班，但沒想到最近公司又開始要辦理產品登記，天啊？我怎麼都逃不了？先前辦理登記讓我工作壓力大到嚴重掉髮，頭也禿了一塊，請救我，我不想禿更多啊～

已不自覺開始查髮片價格的 Claudia

Dear Claudia,

　　等妳查到質好價又優的髮片，記得也告訴我一下，我自己跟我周遭幾個半夜還可在 LINE 上相見的朋友，可都有掉髮的問題呢！回來說說這令妳、也令很多人傷神的登記事，我倒覺得妳這次辦理登記時，妳會不由自主地叫出「不難耶！」。不管是產品登記或是申請任何其他的正式文件，雖然確實會比其他國貿聯繫事來得複雜許多，但是，規定是死的，待妳耐著性子仔細寫完所有的要求，工作其實也就完成了大半，剩下的才是因原廠、因個案而異的狀況，只要妳就每個出現的問題，確實問出答案，這樣一步步下來，離成功完成登記就不遠了！而辦完一次產品登記，妳所得到可是拳拳到肉的硬底子功夫啊！雖然你可能當時沒感覺，但等妳再辦一次產品登記時，妳可就熟門熟路，再怎麼複雜，都不太會有當初第一次接觸時的那種龐大壓力了。其實，遇到我們愈不會的事，只要我們別將太多精神花在焦慮上，直接動手做、耐心學，到最後，一定會讓我們學到飽飽的知識與經驗！所以，請別擔心，請妳放心，妳真的不會禿更多的啦！

278

🔍 單字片語說分明

• certificate [sɚ`tɪfəkɪt]

n. 證書 an official document or record stating that particular facts are true

例 I am now in possession of a Free Sale Certificate duly legalised by the ROC Embassy in to the Holy See.

我現在已拿到了由梵蒂岡的中華民國大使館所核發的自由銷售證明。

certificate [sɚ`tɪfəˌket] Ⓥ 用證書證明

certify [`sɝtəˌfaɪ] Ⓥ

• Power of Attorney

n. 授權書 the authority to act for another person in specified or all legal or financial matters

例 Your company is the only one in Taiwan which has a valid and signed Power of Attorney.

你們是臺灣唯一取得有效且經簽核之授權書的公司。

attorney [ə`tɝnɪ] ⓝ 律師

• authorization [ˌɔθərəˈzeʃən]

n. 授權；認可；批准 official permission to do something

例 Import Permits, if required, will need to be sent separately and we will not go ahead with authorization of order until it is received.

若你們是需要申請進口許可證的話，則請另外發此證給我們，我們收到後才會接著批准處理此訂單。

• authority [əˈθɔrətɪ]

n. 主管機關 an organization or institute that controls something, often a public service

例 The letter needs to be issued by the appropriate competent national authority.

這封信函須由合適的主管機構出具。

其他字義：專家；權威；當局；權力；影響力；說服力

authorize [ˋɔθəˌraɪz] Ⓥ

常見搭配詞

authority figure	with authority
有權勢的人	有說服力
local authority	national authority
當地主管機關	國家主管機關

· registration [ˌrɛdʒɪˋstreʃən]

n. 登記；註冊 the process of recording names or information on an official list

例 I would be obliged if you could confirm timescales at your end for registration.

若您能確認您那兒的登記時程表，我將會非常感激。

register [ˋrɛdʒɪstɚ] Ⓥ 登記；註冊；（儀器）顯示

Ⓝ 正式登記；登記表；（語言學）語域

常見搭配詞

Certificate of Registration	Registered trademark（Ⓡ）
登記證	註冊商標

· distributor [dɪˋstrɪbjətɚ]

n. 經銷商 a company or person that supplies goods to stores

例 The Distributor shall offer technical support to the customers and use its best efforts to promote the sale and distribution of our products.

經銷商須為客戶提供技術支援，盡力銷售與經銷我們的產品。

distribute [dɪˋstrɪbjʊt] Ⓥ 分發；配送；散佈

PART
1
PART
2
PART
3
PART
4
PART
5
PART
6
PART
7

· permit [`pɝˋmɪt]

n. 允許；許可 an official document that gives you permission to do something

例 We will have the product ready when your import permit is available.

等你們拿到進口許可證時，我們的貨也會準備好。

permit [pɚˋmɪt] Ⓥ 允許；許可；使成可能

permission [pɚˋmɪʃən] Ⓝ   permissible [pɚˋmɪsəbl] adj

常見搭配詞

a work/ travel/ export permit   weather permission

工作證／旅遊證／出口證   如果天氣允許的話

ask sb's permission   get/ obtain sb's permission

徵求同意   取得許可

· statement [`stetmənt]

n. 聲明 a written or spoken announcement on an important subject that someone makes in public

例 We can offer the following statement about our product. Please let me know if this is sufficient.

我們可提供與產品有關的聲明書，請告知是否這樣就已足夠了。

其他字義：證詞；對帳單；說明

state [stet] Ⓥ 陳述；說明；規定

常見搭配詞

make/ issue/ deliver/ give/ release a statement 提出聲明

joint statement   state a fact/ opinion   state the obvious

聯合聲明   陳述事實／意見   陳述顯而易見的事

• representative [rɛprɪ`zɛntətɪv]

n. 代表；代理人 someone who has been chosen or elected by a person or group to vote, give opinions, or make decisions for them

例 Please contact our Customer Service Representative for more information.

如您需要更多的資訊，請與我們的客服代表連絡。

其他字義：adj. 典型的；代表性的

represent [ˌrɛprɪ`zɛnt] v. 代表；表示；象徵；描繪；具體呈現

representation [ˌrɛprɪzɛn`teʃən] n.

常見搭配詞

truly/ genuinely representative 有真正的代表性的

a representative sample/ selection 有代表性的樣本

• entitle [ɪn`taɪtl]

v. 給…權力（或資格）to give someone the right to do something

例 This membership entitles the holder to enjoy all the benefits.

具會員身分可讓會員有資格享受所有的福利。

其他字義：給（書、詩歌、樂曲）題名；給…命名

entitlement [ɪn`taɪtlmənt] n.（接受某物或做某事的）權利、資格

· grant [grænt]

v. 同意；准許 to allow someone to have or do what they want

例 If permission is granted, I agree to abide by the above conditions.

若是取得了核准，則我可同意遵照上述的條件。

其他字義：承認

grant [grænt] n 資助；撥款；補助金

常見搭配詞

take sb/sth for granted 視（某人的存在和幫助／某事）為理所當然

take it for granted (that)   I grant you (that)

理所當然地認為   我承認

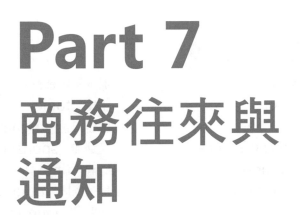

Part 7
商務往來與通知

Unit 1　會議與展覽
Unit 2　通知

Unit 01

會議與展覽

單字片語一家親

展覽、會議大代誌

展覽　exhibition

國際會議
international conference

博覽會　exposition

年會　annual meeting

聯合年會
joint annual conference

經銷商會議
distributor meeting

研討會　seminar, workshop

線上研討會　webinar

會議；講習會　session

演講　speech

專題演講　keynote speech

論文發表　paper presentation

海報發表　poster presentation

教育訓練　training

銷售訓練課程
sales training course

技術訓練課程
technical training course

展覽基本資料

主題　topic

主辦單位　organizer

贊助商　sponsor

時間　time

展覽期間　duration

地點　venue, location

展覽館　exhibition hall

登記；註冊　registration

會議流程怎麼走？

初定議程　preliminary agenda

暫定計畫　tentative schedule

會議程序　conference program

展覽會上會提供⋯

文宣　marketing materials
　　　　literature

型錄　catalog

主題型錄　topical catalog

促銷品　promotional items

贈品　giveaways

免費樣品　free sample

折價券　coupon

促銷代碼　promo code

抽獎　drawing

句型

句型 1 邀請參加盛會！

I would like to invite sb to attend + event + which will take place in + place + on + time.

例 I would like to invite you to attend the exhibition opening which will take place in the hall on the 23rd of November.

我想要邀請您參加 11 月 23 日在大廳舉行的展覽開幕式。

句型 2 好好準備準備！

We also would like to request some additional information from you so that we can prepare for sth.

例 We also would like to request some additional information from you so that we can prepare for your visit.

我們也想跟您要些資料，好讓我們能為您的來訪先做些準備。

句型 3 議程來了！

Here is the preliminary agenda for sth.

例 Here is the preliminary agenda for our annual meeting.

在此送上我們年會的初定議程。

句型 4 今日快報！

This announcement is to notify you of sth.

例 This announcement is to notify you of the next scheduled meeting.

這份通知是要來告訴您預計舉行的下一場會議。

句型 5 歡迎來攤位一遊！

You are warmly welcome to visit us on our booth # xxx at any time.

例 You are warmly welcome to visit us on our booth # 28 at any time.

歡迎您隨時來我們的 28 號展覽攤位參觀。

PART 1

PART 2

PART 3

PART 4

PART 5

PART 6

PART 7

安排會議 E-mail | Scheduling a Meeting

代理商會議通知 E-mail 範例 · 會議主旨、可提出個別會面

Dear Sarah,

I would like to invite Hope Tech to attend our 2015 Distributor Meeting which will take place in Ann Arbor, MI on June11th - 12th. Attached is the tentative schedule for your review and for travel planning purposes.

Please note that this distributor meeting will focus primarily on sales & marketing tools and strategies to promote our products. We will also offer an opportunity for you to request a 30-minute individual meeting with our staff to review your sales performance and discuss any questions or concerns you may have. If you would like to request an individual meeting, please let me know 3 choices for your time slot so that we can add you to the schedule.

Please let us know if you have any questions or concerns. We hope to see you in June!

Sincerely,
Jeffrey Fisher
Calibration Corp.

單字 ShabuShabu 一小補小補！
☑ tentative [ˋtɛntətɪv] (adj.) 暫時性的
☑ primarily [praɪˋmɛrəlɪ] (adv.) 主要地
☑ time slot 時段
☑ calibration [ˌkælɪˋbreʃən] 校準

PART
1
PART
2
PART
3
PART
4
PART
5
PART
6
PART
7

代理商會議通知 **E-mail** 範例 中文

莎拉，您好，

我想要來邀請賀普科技參加我們將在六月十一日、十二日舉行的 2015 年經銷商會議，地點設於邁阿密安娜堡，在此附上暫定的行程資料供您參考，也供您安排出國行程。

請注意這次經銷商會議的主題將定在推廣我們產品之銷售與行銷的工具與策略，我們亦提供了可讓您與我們人員個別舉行 30 分鐘會議的機會，以回顧審視您的業績表現，您也可提出任何問題與議題來討論。若是您要與跟我們個別開會，請給出三個時段供我們選擇，這樣我們就可將您排進會議計畫中。

若是您有任何的問題與考量，就請告訴我們，我們期待能在六月跟您會面。

謹上
傑佛瑞·費雪
佳立博公司

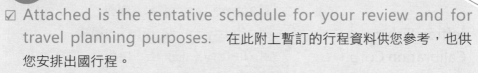

Check 好句

☑ Attached is the tentative schedule for your review and for travel planning purposes. 在此附上暫訂的行程資料供您參考，也供您安排出國行程。

☑ This distributor meeting will focus primarily on sales & marketing. 這次經銷商會議的主題將定在銷售與行銷。

Dear Sarah,

We are excited for next week's Distributor Meeting and look forward to your arrival! For your reference, I have attached the most current agenda for our sessions on Thursday, June 11th and Friday, June 12th.

We also would like to request some additional information from you so that we can prepare for your arrival:

- Please tell us your title. We'll have it appeared on your name tag.
- Do you have any food allergies? Any dietary restrictions?

If you have questions or concerns, please do not hesitate to contact us.

Travel safely. See you next week!

Sincerely,

Jeffrey Fisher
Calibration Corp.

單字 ShabuShabu 一小補小補！
☑ name tag   名牌；胸牌
☑ allergy [ˋælə‧dʒɪ] (n.)   過敏
☑ dietary [ˋdaɪə‚tɛrɪ] (adj.)   飲食的

PART
1

PART
2

PART
3

PART
4

PART
5

PART
6

PART
7

會議前提醒 E-mail 範例 中文

莎拉，您好，

　　下星期就要開經銷商會議了，我們很興奮，也很期待您的到訪！在此附上六月十一日（四）、十二日（五）所舉行之會議的最新議程供您參考。

　　我們也想跟您索取些資料，好讓我們能為您的到訪做準備：

- 請告知您的頭銜，我們會將其列在你的名牌上。
- 您對任何食物會過敏嗎？飲食上有什麼限制嗎？

　　若是您有任何的問題與考量，就請儘管與我們聯絡。

　　祝您旅途平安，期待下星期與您相見！

謹上

傑佛瑞・費雪
佳立博公司

Check 好句

☑ We are excited for next week's Distributor Meeting and look forward to your arrival!　下星期就要開經銷商會議了，我們很興奮，也很期待您的到訪！

☑ Travel safely. See you next week.　祝您旅途平安，期待下星期與您相見！

安排教育訓練課程 E-mail | Arranging a Training Course

教育訓練通知 E-mail 範例 · 會議主旨、可提出個別會面

Dear Jane,

Crystal Technologies is pleased to welcome you to our 3-day distributor sales training that will take place at the Holiday Inn Frankfurt, Germany the 23rd -25th of November, 2015. The training will include seminars on various research areas where Crystal Technologies products are used. The seminars are followed by discussions among the participants in smaller groups. The discussion outcome is presented and discussed in full group. Participants are asked to give a presentation of approximately 5 minutes about their respective market, and also the target customer base, product portfolio and the marketing strategies of Crystal Technologies products. Please be prepared for this presentation!

Here is the preliminary agenda for Crystal Technologies Distributor Sales Training 2015, Frankfurt 23-25 November. If you have any other questions, don't hesitate to contact me.

Best Regards,

Steven Clooney
Crystal Technologies

單字 ShabuShabu 一小補小補！
☑ participant [parˋtɪsəpənt] (n.) 參與人員
☑ respective [rɪˋspɛktɪv] (adj.) 個別的
☑ portfolio [portˋfolɪ͵o] (n.) 組合

教育訓練通知 **E-mail** 範例 中文

珍，您好，

　　水晶科技即將舉行為期三天的經銷商銷售教育訓練，我們很高興邀請您前來參加，地點設在德國法蘭克福的假日飯店，時間為 2015 年十一月二十三日～二十五日。這次的教育訓練將會有幾場的研討會，以水晶科技產品應用的各個不同研究領域為主題。研討會結束後，參加的成員就會分組討論，各小組的結果將以簡報的方式來跟所有的人員分享，並一起討論。參加的每個人也要提出簡報，時間約五分鐘，介紹各自代表的市場，並說明水晶科技產品的目標客群、產品組合，還有行銷策略，還請您為這場簡報做好準備。

　　在此附上水晶科技將於十一月二十三日～二十五日在法蘭克福舉行之 2015 經銷商銷售教育訓練的初定議程，若您有任何其他的問題，請儘管與我聯絡。

祝好

史帝文 · 克魯尼
水晶科技

Check 好句

☑ We are pleased to welcome you to our 3-day distributor sales training. 我們很高興邀請您參加我們為期三天的經銷商銷售教育訓練。
☑ The seminars are followed by discussions among the participants in smaller groups. 研討會結束後，參加的成員就會分組討論。

線上研討會通知 E-mail 範例 ・時段選擇

Dear Christine,

This year, Grandeur Tech will be holding quarterly webinars for all of our distributors. <u>This announcement is to notify you of the first one to be held in two weeks.</u> Below are the date and time each session will be held:

Thursday, 5th of March   5:00 pm (PST)
Friday, 6th of March   8:00 am (PST)

The webinars will be held via www.joinwebinar.com and last approximately 1 hour with 10-15 minutes for questions and answers. The presentation will be held by Kevin Cantley, Ph.D., Grandeur Tech's Technical Manager. The topic for this webinar is 5X1-Power Series".

Please let me know which session you will attend and how many people from your company will call in. We do have a limited number of participants, so I encourage you to call in as a group.

Kind Regards,

Joyce Perkins
Grandeur Tech

單字 ShabuShabu 一小補小補！
☑ grandeur [ˋgrændʒɚ] (n.) 宏偉
☑ quarterly [ˋkwɔrtɚlɪ] (adj.) 每季的
☑ encourage [ɪnˋkɝɪdʒ] (v.) 鼓勵

PART
1

PART
2

PART
3

PART
4

PART
5

PART
6

PART
7

線上研討會通知 **E-mail** 範例 中文

克莉絲汀，您好，

　　今年宏大科技將為我們所有的經銷商舉辦每季的線上研討會，這份通知是要來告訴您第一場研討會將在兩週後開跑。每個場次舉行的日期與時間如下：

　　三月五日，星期四，下午五點（太平洋時區）
　　三月六日，星期五，上午八點（太平洋時區）

　　線上研討會將透過 www.joinwebinar.com 舉行，時間約一小時，另開放十到十五分鐘的提問時間。此次簡報將由宏大科技的技術經理 - 凱文・坎特利博士報告，研討會主題為「X1-Power 系列」。

　　請告知您欲參加哪一個場次，亦請告知您公司將有多少人會上線參加。因為參與人數有限制，所以建議您以團體方式上線。

祝好
喬伊絲・柏金斯
宏大科技

Check 必 好句

☑ Below are the date and time each session will be held.　每場研討會舉行的日期與時間如下。

☑ The webinars will last approximately 1 hour with 10-15 minutes for Q&A.　線上研討會舉行時間約一小時，另開放十到十五分鐘的提問時間。

參加展覽 E-mail | Attending an Exhibition

邀請參加展覽 E-mail 範例 · 可約定會面時間、還可參加另一個展覽

Dear Jeannie,

Horizon Science Inc. will attend BCS 2015 15th Annual Meeting and Exhibition in Orlando on July 10-12. If you or a representative of your company is planning to attend, we would heartily welcome the opportunity to meet with you. Please send your e-mail to marketing@horizonscience.com to book a meeting. In case you don't have the opportunity to book a meeting in advance, you are warmly welcome to visit us on our booth #3131 at any time.

You are also most welcome to attend our Industry Workshop which takes place the day before the full meeting and exhibition begin. The subject is "Celebrating 20 Years of Achievement in Cardiac Disease Diagnostics". You can visit our website for details.

See you soon in Orlando!

Best regards,

Alisa Taylor
Horizon Science Inc.

單字 ShabuShabu 一小補小補！
horizon [hə`raɪzn] (n.) 地平線
industry [`ɪndəstrɪ] (n.) 產業
celebrate [`sɛləˌbret] (v.) 慶祝
cardiac [`kɑrdɪˌæk] (adj.) 心臟的

邀請參加展覽 E-mail 範例　**中文**

吉妮，您好，

　　地平線科學公司將會參加於七月十日～十二日舉行的 BCS 2015 年第十五屆年會與展覽，地點在奧蘭多，若您或您公司的代表有計畫前來參加，我們竭誠歡迎並期待與您會面。請發 e-mail 至 marketing@horizonscience.com，以預約會面的時間。若是您沒能事先預約，我們亦隨時歡迎您前來3131 號展覽攤位，與我們碰面。

　　我們也很歡迎您參加我們的產業研討會，此研討會將於年會與展覽開始的前一天舉行，主題為「慶祝心臟疾病診斷之二十年成果」。如需瞭解相關明細，請查閱我們的網站。

　　期待儘快在奧蘭多與您相會。

祝好

艾莉莎・泰勒
地平線科學公司

Check 好句

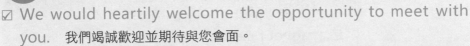

☑ We would heartily welcome the opportunity to meet with you.　我們竭誠歡迎並期待與您會面。

☑ You can visit our website for details.　如需瞭解相關明細，請查閱我們的網站。

提供參展支援 E-mail 範例 · 提供文宣、詢問展覽相關訊息

Dear John,

We would be happy to work with you to provide updated literature, catalogs, and some promotional items (including magnets, pens and stuffed underwater animals) to distribute to visitors at the upcoming JADE Exposition in October. We can also provide our environmental shopping bags for you to give to potential customers. Attached is the image of the bag. It is a plastic bag that features an underwater world.

Please send me the link to the exposition so that our Technical Writers can review its related topics to make sure that the appropriate literature is provided.

We will be able to send 2 boxes of bags along with the other promotional materials to you no later than October 2nd.

If you have any additional questions or concerns, please do not hesitate to contact me. Thank you!

Sincerely,

Betty Rose
Explorer Diving Equipment Co.

單字 ShabuShabu 一小補小補！
☑ magnet [`mægnɪt] 磁鐵
☑ stuffed [stʌft] (adj.) 填充的
☑ environmental [ɪnˌvaɪrən`mɛntl] (adj.) 環保的

提供參展支援 E-mail 範例 中文

約翰,您好,

我們很樂意配合提供最新的文宣、型錄,以及一些促銷禮品(包括磁鐵、筆、海底動物填充玩偶),讓您在十月舉行的 JADE 博覽會上發送給參觀的人,我們也可提供免費的環保購物袋,讓您送給潛在客戶。在此附上袋子的圖片,它是塑膠材質的袋子,上面有海底世界的圖案。

請提供給我這個展覽的網站連結,好讓我們的技術文件工程師能瀏覽一下相關的主題,確保提供適合的文宣資料給您。

我們最晚會在十月二日之前,將兩箱的袋子連同其他促銷用的物品寄給您。

若是您有任何其他的問題或考量,請儘管直接與我連絡,謝謝!

謹上
貝蒂・羅斯
探索家潛水設備公司

Check 好句

☑ We would be happy to work with you to provide some promotional items to distribute to visitors. 我們很樂意配合提供一些促銷禮品,讓您發送給參觀的人。

☑ Please send me the link to the exposition so that we can review its related topics. 請給我這個展覽的網站連結,好讓我們能瀏覽一下相關的主題。

☎ 電話對話

電話討論來訪行程 範例

人物介紹

Paul

保羅，經銷商業務主管，做事仔細，認真確認細節！

Ariel

艾芮兒，原廠行銷部門主管，做事積極，目標訂得也積極！

Ariel: Hello, this is Ariel.

Paul: Hi, Ariel. This is Paul Chang. I'm calling to confirm your flight itinerary.

Ariel: I just got the details! I will fly from Hong Kong in Flight number CX 0474 with Cathay Pacific. The estimated landing time is at 10:20 on the Wednesday morning of Jan. 28th.

Paul: Got it! I'll pick you up at the airport and bring you to our office.

Ariel: Great! Thanks. For your information, at the meeting, I'll provide you with a summary presentation for the year 2014.

Paul: Good. We've prepared our presentation on the 2014 sales and also the 2015 marketing strategies.

Ariel: Wonderful! Also please share with me any promotional campaigns that you plan to run through out this year, so we could know how to best help you.

Paul: Not a problem!

Ariel: We have set an aggressive 2015 worldwide growth target at 35%. I'd like to know what you will need from us to achieve this.

Paul: That target is quite high! We'll evaluate first and let's discuss this further when we meet.

Ariel: Sure thing! I'm looking forward to meeting with you soon!

PART
1

PART
2

PART
3

PART
4

PART
5

PART
6

PART
7

電話討論來訪行程 範例 中文

艾芮兒：你好，我是艾芮兒。

保　羅：嗨，艾芮兒，我是張保羅，我打來是要跟妳確認妳的班機行程。

艾芮兒：我剛拿到資料明細呢！我會搭乘國泰航空 CX 0474 班機，從香港飛臺灣，預計一月二十八日星期三早上 10:20 抵達。

保　羅：了解！我會去機場接您到我們公司。

艾芮兒：太棒了！謝謝你。先讓你知道一下，開會的時候我會做個簡報，回顧一下 2014 年的業績總結。

保　羅：好的，我們也已經準備了我們的簡報，要來跟妳說明 2014 年的銷售狀況，以及 2015 年的行銷策略。

艾芮兒：太好了！另外也請告訴我你們今年一整年所計畫的促銷活動，這樣我們也好盡力來協助你們。

保　羅：沒問題！

艾芮兒：我們已經設定了 2015 年全球成長目標要達到 35%，所以我也想知道你們會需要我們怎麼協助，好讓你們能達到這個目標。

保　羅：這個目標很高耶！我們會先評估看看，等開會的時候再來讓我們好好討論一下。

艾芮兒：當然！期待儘快跟你們開會囉！

電話英文短句一好說、說好、說得好！

・I'm calling to confirm your flight itinerary.　我打來是要跟您確認您的班機行程。

・Sure thing!　當然！

國貿知識補給站　會議與展覽的對「外」溝通

　　平時面對著電腦與國外連繫，總是心中呼著對方的名，腦海中想像著國外在看到你的 e-mail 後會怎麼回。遇到重要或急切的事，就是電話連絡的現「聲」時刻了。而拜科技之賜，現在還可選用視訊的現「影」服務，為你與國外搭起友誼的橋樑！不過，有聲有影還不夠，每年總有好些個會議與展覽，非得要現「身」才能談得盡興、完成使命。這樣的聚會與場合，可都是各個參與的公司呈現業績表現與展現能耐的年度大事呢！在此，我們就來看看這些大事在聯繫與溝通上有什麼該注意的地方囉！

會議

　　會議的時間可是頭一個要先討論與設定的要點，若是多方參與的經銷商會議或訓練課程，原廠說了就算，請各方人馬回報參加與否即可。若是一對一的公司會議，提議的一方應提早通知對方，提出個大致的時間或幾個日期，來供對方選擇。在敲定日期後就要來敲定議程，將何時開會、何時結束寫得一清二楚。而到了開會前幾天以及前一天時，請再發個 e-mail 提醒對方。這是讓對方不能忘、不敢忘或若記錯時間還來得及導正的小舉動，也是讓你不會遇上等不到人的麻煩情境。

　　再來，開會的人員有誰、職務為何，為原廠安排之客戶拜訪的機構名與人名資料，也都是需要告知對方的必要資訊。如此一來，與會人員對於對談的對象都能有概括的了解，彼此在簡報、討論的涵蓋主題與深度方面，也都可預先設計與安排。

　　明白地定出會議討論主題並事先告訴對方，是確保會議能否有效進行的關鍵。你大可在會議時邊談邊提你想到的主題，但這樣一來，第

一次接招的對方也就有可能在會議中跟你說「我查了之後再用 e-mail 回你」，而這樣的對談結果則是浪費了面對面會談的寶貴時間，也會讓問題立馬處理、提早解決的大好機會溜走！

在會議進行時，請確實做出會議紀錄，會後則就此會議紀錄所記下的代辦事項，逐一進行、完成，讓雙方難得的會議不光是說說而已，而是說了都有處理，處理了就有改善，就有進展，這樣才是確實發揮了會議的效益！

展覽

展覽可是提高企業曝光度、強化品牌印象、吸納新客戶、擴增客戶量的好機會！參展企業在花了錢、搶得好攤位之後，接下來就要好好聯繫攤位設計與活動內容了。就代理商而言，展覽可替國外原廠打知名度，而代理商會要求原廠提供的協助，從成本分攤、抽獎獎品、折扣券、樣品、贈品到文宣都有。那怎麼樣才能取得原廠的最大協助呢？說穿了就是多說點、說透徹點，將展覽的重要性、規模、預期效益、估計參觀人數、競爭廠牌的參展狀況逐一說明，接著再談談企業的活動規畫與設計，告訴原廠會有哪些地方需要原廠協助，以及如何協助。談定之後，就請盯緊時程，確保原廠所要提供的物品與文宣可準時在展覽前準備就緒。

等到展覽結束，因為對國外不再有需要協助的地方，所以常有對原廠就此沒有任何後續說明的狀況發生。實際上，辦了個活動，事前既然有預估效益，結束後，當然也就該有實際效益的評估。在展覽後趁熱所做的成效分析，是參展企業檢討不足、調整下次參展策略的最佳時機，而將此分析結果回報給原廠，則是讓原廠得悉全貌、調整下次參展配合與協助的最好根據，也是讓原廠肯定代理商辦事能力的又一事例！

會議與展覽都是企業的大事，大事好好辦、仔細辦、有頭有尾，將來成就的就會是大事業了！

《Dear Amy》時間

Dear Amy,

我是個英文業務秘書，每天都要寫好幾十封 e-mail 給原廠，照理說，我應該很懂得與國外溝通，但每次一看到國外原廠發來 e-mail，說要來臺灣開會，我就會整個心一下子揪在一起，開始緊張起來，一直要到原廠人來了、人走了，我才會洩氣式地整個放鬆。每次都是這樣，都沒有一次比一次不緊張，我自己都快受不了了，而且，緊張的我在面對原廠人員時，臉上肌肉都快控制不住地抖動起來了，怎麼辦？我是不是有 xenophobia，有外國人恐懼症啊？

長得大隻，但遇到外國人就整個縮起來的 Helen

Dear Helen,

請妳放心，每個要與國外開會的人，心裡多多少少都會緊張的哩！但這分緊張並沒有不好，因為重視，自然心裡就會緊張，而因應的方式無他，就是要求自己多多準備，盡量讓自己在開會時每一件事都聽得懂，都知道要點在哪。事前的準備要如何做呢？國外原廠來臺開會都會有設定主題與方向，所以我們可依已知的訊息與手邊的資料，來沙盤推演一下！國外業務主管來，一定談業績、談數字、談市場狀況、談競爭，請自擬一套不太短的回答與市場分析，要求自己練習用英文說出，練習時若有哪個字彙不確定，馬上查出來，記下來。做過了這樣的動作，妳會發現在真正開會討論時，或許有更多種數據與事件冒了出來，但是大方面都在，並不會有太大的歧異！要是妳覺得光是頭腦想，嘴巴練習，還無法讓妳放心，那就請將妳的模擬會議對話寫下來，多寫幾句，多練習幾遍，就可多些備戰功力，多些應戰信心了！開會偶爾還是會碰到我們一無所悉的主題，這時妳可能容易聽了會恍神，會落了一大段的不知所云，碰到這樣的情況怎麼辦呢？請拔尖了耳朵聽！藉機訓練自己找言談中蛛絲馬跡的能力，而當妳對所做的工作內容懂得愈多，妳就愈有能力找線索，愈能更快跟上討論的步調，這樣一來，妳就不會整個人縮起來，而是會愈坐愈挺，盡情融入開會的情境呢！

單字片語說分明

- annual [ˋænjʊəl]

  adj. 一年一次的 happening once a year

  例 Our annual clearance sale is on! All the auction items will be offered 60% off discount!

  我們的年度清倉大拍賣來了！所有拍賣品項全都下殺 4 折！

  其他字義：年度的；全年的

  比較比較：biannual [baɪˋænjʊəl] adj 半年一次的

  biennial [baɪˋɛnɪəl] adj 兩年一次的

  常見搭配詞

  an annual meeting 年會　　　　　an annual conference 年度會議

  an annual salary 年薪　　　　　　an annual total 全年總額

- keynote [ˋkiˌnot]

  n.（演說等的）主旨；基調 the most important feature of something

  例 Our CEO is invited to deliver a keynote speech at the conference.

  我們的執行長獲邀在會議中發表主題演講。

  其他字義：（音樂）主音

  keynote [ˋkiˌnot] v 以…做為主旨提出

  常見搭配詞

  keynote speech/ address　　　　keynote speaker

  主題演說　　　　　　　　　　　　主題講者

305

• presentation [ˌprizɛnˋteʃən]

n. 簡報 a formal talk in which you describe or explain something to a group of people

例 You could take a look at Facebook for pictures of the event and also watch our Product Manager's presentation at our Youtube channel.

您可以在臉書上看看這次活動的照片，也可以在 Youtube 上看我們產品經理所做的簡報。

其他字義：贈送；授予；提出；呈現；演出

present [prɪˋzɛnt] Ⓥ 贈送；導致；提出；呈現；演出

presentable [prɪˋzɛntəbl] adj 像樣的；體面的

present [ˋprɛznt] ⓝ 禮物；贈品/adj 出席的；現在的

presence [ˋprɛzns] ⓝ 出席；風度；風範

常見搭配詞

make/ give a presentation (n.)	present (v.) a problem/difficulty
做簡報	會有困難
make yourself presentable (adj.)	make your presence (n.)
把自己打扮得像樣些	使自己令人注目
in sb's presence (n.)	presence (n.) of mind
有某人在場	鎮定；沉著
at present (adj.)	for the present (adj.)
目前	暫時；就目前來說

• organizer [ˋɔrgəˌnaɪzɚ]

n. 主辦單位；主辦者 someone who makes all the arrangements for an event or activity, especially as a job

例 We are the leading organizer of hands-on training courses for SAP ERP software.

我們是 SAP 企業資源規劃軟體實用訓練課程的主要主辦公司。

organize [ˋɔrgəˌnaɪz] Ⓥ 組織；安排；使有調理；組織工會

organized [ˋɔrgənˌaɪzd] ⓐⓓⓙ 有組織的；有條理的

organization [ˌɔrgənəˋzeʃən] ⓝ 組織；機構；體制；結構

organizational [ˌɔrgənaɪˋzeʃənəl] ⓐⓓⓙ 組織的

常見搭配詞

highly organized 組織得非常嚴密　　organizational skills 組織才能

a voluntary/ charity/ aid organization 志願／慈善／援助機構

a non-governmental organization (NGO) 非政府組織

a non-profit organization (NPO) 非營利組織

・sponsor [ˋspɑnsɚ]

n. 贊助商 a person or business that pays money to support an event, a television or radio program, a website etc as a way to advertise their products or services

例 We are pleased to announce that our company is a Platinum sponsor of this international conference.

我們很高興要來通知您，我們公司是這一個國際會議的白金級贊助商。

其他字義：舉辦者；支持者；提案人；擔保人

sponsorship [ˋspɑnsɚˌʃɪp] 資助金；贊助費

・duration [djʊˋreʃən]

n. 持續期間 the period of time during which something continues to happen or exist

例 Please provide the name and contact details of the customer and also duration of discount, so I can issue a discount code.

請提供客戶名與其連絡明細，以及折扣的有效期，有了這樣的訊息，我們就可發出折扣代碼給您。

其他字義：舉辦者；支持者；提案人；擔保人

durable [`djʊrəb!] adj. 耐用的；持久的

durability [ˌdjʊrə`bɪlətɪ] n. 耐久性

常見搭配詞

in duration　持續　　　　　　for the duration　在此期間

a durable solution　　　　　 durable goods

一勞永逸的解決方案　　　　　耐用品

• preliminary [prɪ`lɪməˌnɛrɪ]

adj. 初步的；預備性的 coming before the main or most important part of something

例 Please note that this assay provides only a preliminary analytical test result.

請注意這個試劑只可提供初步的分析檢測結果。

preliminary [prɪ`lɪməˌnɛrɪ] n. 序言；開場白；引介；準備工作；預賽（用複數）

• agenda [ə`dʒɛndə]

n. 議程 a list of things that people will discuss at a meeting

例 All confirmed delegates will receive the agenda in the next few days.

所有確定參加的代表人員，都會在這幾天內收到議程資料。

其他字義：代辦事項

常見搭配詞

high on the agenda　　　　　at the top of my agenda

工作的重要性高　　　　　　　我的首要考量

be on the agenda　　　　　　the political agenda

列入議程　　　　　　　　　　政治議程

· literature [`lɪtərətʃɚ]

n. 印刷品；文宣 books or other printed information about a subject

例 Here are some examples of marketing literature we have designed.

這些是我們設計出來的幾個行銷文宣樣本。

其他字義：文學；文學作品；文獻；學術著作

NOTES

Unit 02

通知

單字片語一家親

組織變革大代誌！

合併	merger
收購	acquisition
接管	takeover
加入	join
同盟	alliance
分公司	branch company
	affiliated company
子公司	subsidiary

一定要說有這些好處…

提升效率 increase efficiency

營運上更有效率

operate more efficiently

擴充產能	expand capability
發揮綜效	develop synergy
精簡流程	streamline process
降低成本	reduce/cut/save costs
擴增銷售基礎	increase sales base
互惠	mutual benefit
多元化經營	diversify
重整	reconstruct
鞏固	consolidate

變革的影響…

維持不變 remain unchanged

營運如常	operate normally
小小調整	adjust slightly
大大調整	adjust significantly
新產品型號	new product code
新版面編排	new layout
新格式	new format
新銀行帳戶	new bank account

新的一年要來了！

接近了	approaching
快來了	around the corner

年度回顧

成功	success
繁榮；昌盛	prosperity
成就	accomplishment
簡短回顧	brief review
更新	update

致謝

表達感謝 express gratitude

感謝 thank, appreciate,
be grateful for,
be obliged to, acknowledge

支持	support
合作	cooperation
協助	assistance

句型

句型 1 被併了！

As you may know, A has recently been acquired by B.

例 As you may know, the company has recently been acquired by a multinational corporation.

或許您已知道，這家公司最近被一家跨國企業併購了。

句型 2 合體中！

We are in the process of combining sth.

例 We are in the process of combining them into a single and more dynamic system.

我們目前正在將它們整合成單一且更具機動性的系統。

句型 3 話說影響程度…

It is anticipated that the merger will have + a(n) + adj. + effect on current and future operations.

例 It is anticipated that the merger will have a significant effect on current and future situations.

此併購預計會對目前與未來營運上造成很大的影響。

句型 4 新年新提醒！

As we move into the new year, I would like to send you a few reminders and update you on sth.

例 As we move into the new year, I would like to send you a few reminders and update you on our latest developments.

在我們邁向新年度之際，我想要提醒您幾件事，也跟您說說我們最新的發展狀況。

句型 5 回顧一下！

We'd like to take this opportunity to give you a brief review of sth.

例 We'd like to take this opportunity to give you a brief review of the current situation.

藉著這個機會，我們要來跟您簡短回顧一下目前的狀況。

311

國貿英語 溝通術
Master English　Communication for International Trade

組織變動通知 E-mail | Notifying of Organizational Change

併購通知 E-mail 範例 1 · 小調型號與格式、產品品質不變

Dear Debby,

As you may know, Calcium Corp. has recently been acquired by its long time supplier, Biosynthesis. This acquisition will allow Calcium Corp. additional efficiencies, including expanded customer service capabilities and faster shipping time.

We are in the process of combining our two companies into one team that will be ideally positioned to meet the challenges and opportunities of the healthcare market in the future. Our primary consideration is to ensure our customers are fully supported!

We will be slightly adjusting product codes and also the layout of our Certificate of Analysis to match Biosynthesis's format. Please rest assured that the actual material you will be receiving has not changed. Thank you for your continued business and support!

Kind regards,

Tracey Klein
Customer Service Manager

單字 ShabuShabu 一小補小補！
☑ calcium [ˈkælsɪəm] (n.)　鈣
☑ challenge [ˈtʃælɪndʒ] (n.)　挑戰
☑ opportunity [ˌɑpəˈtjunətɪ] (n.)　機會

併購通知 E-mail 範例 1 中文

黛比，您好，

　　或許您已經知道鈣業公司最近已被我們長久以來合作的供應商—拜爾合成所收購，此併購可讓鈣業公司增加額外的功效，包括擴充客戶服務能力，以及加快出貨的速度。

　　我們目前正在將兩家公司整合成一個團隊，期待找出最佳的定位，以能在保健市場中迎接挑戰、抓住機會，我們主要的考量就是要確保能夠全力支援客戶所需！

　　我們將會略微調整產品的編碼與分析報告書的排版，以配合拜爾合成的格式要求。關於產品本身，請您放心，您所收到的貨品將和先前完全一樣。謝謝您持續的生意往來與支持！

祝好

翠西・克萊
客戶服務經理

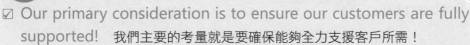

必Check 好句

☑ Our primary consideration is to ensure our customers are fully supported!　我們主要的考量就是要確保能夠全力支援客戶所需！

☑ Thank you for your continued business and support!　謝謝您持續的生意往來與支持！

併購通知 E-mail 範例 2 · 一切幾乎不變、併購前後付款注意事項

Dear Lisa,

We wrote to you in February to advise you of some changes that were about to happen at SBI. These changes have now taken place and this e-mail includes important updated information relating to these changes. On 1 April 2015 the SBI officially joined the HMI Group. These two organizations have already worked closely together for some time. The work to bring the two organizations together has now been completed.

SBI and its staff will continue to work as normal and it is anticipated that the merger will have a little effect on current and future operations. Invoices which were issued up till 31 March 2015 should be paid to SBI's previous bank account. Invoices which are issued from 1 April 2015 will show new bank account and bear the new SBI logo and the logo of the HMI Group.

If you have any further questions relating to this merger, please contact your normal SBI counterpart. Thanks.

Best regards,

Kevin Dobbs
SBI

單字 ShabuShabu 一小補小補！
☑ take place　發生
☑ officially [əˋfɪʃəlɪ] (adv.)　正式地
☑ counterpart [ˋkaʊntɚˏpɑrt]　對應的人（或物）

併購通知 E-mail 範例 2 中文

麗莎，您好，

　　二月時我們有先寫信給您，通知您 SBI 內部將會出現一些變化，而這些變化目前已經發生，此 e-mail 中就包含了有關這些變化的重要更新訊息。在 2015 年四月一日這一天，SBI 正式加入 HMI 集團。這兩個組織已密切運作了一段時間，而現在將其合併在一起的整合工作也已經完成。

　　SBI 與所有人員將會繼續如常作業，我們預期此併購並不會為現在與往後的營運造成什麼影響。在發票部分，2015 年三月三十一日之前所開立的發票，應匯給 SBI 先前的銀行帳戶，而在 2015 年四月一日之後開立的發票，上頭會有新的銀行帳號，也會有 SBI 的新商標，以及 HMI 集團的商標。

　　若您對此併購案還有其他的問題，請與您平時連絡的 SBI 人員接洽，謝謝。

祝好
凱文 · 道博
SBI

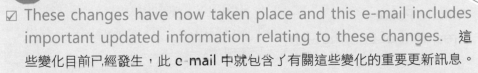

必 Check 好句

☑ These changes have now taken place and this e-mail includes important updated information relating to these changes.　這些變化目前已經發生，此 e-mail 中就包含了有關這些變化的重要更新訊息。

☑ The work to bring the two organizations together has now been completed.　將兩個組織併在一起的整合工作，現在已經完成了。

✉ 感謝與通知 E-mail | Gratitude & Notification

年終感謝與通知 E-mail 範例 **1** ‧ 假期前後出貨計畫、新網站

Dear Distributor partner,

First of all, I want to express our gratitude for your hard work this year! EnzymeTech continues to steadily grow and that is due in large part to your help with making EnzymeTech a globally recognized brand. <u>As we move into the new year, I would like to send you a reminder and update you on an exciting change to look forward to in the coming months.</u>

Holiday Shipping Schedule

December 24 & 25, 2014 + January 1, 2015: CLOSED

We will be open on December 22nd, 26th & 29th, 2014, and only ship orders by request.

New website

We will be launching a brand-new website next month. We will also have a fully updated catalog that will be ready in early 2015.

Thanks again for all of your hard work. We are so excited for 2015 and our continued partnership. All of us at EnzymeTech wish you a very happy and healthy holiday season!

Best regards,
Mikale Pokora
President of EnzymeTech

單字 ShabuShabu 一小補小補！
- ☑ steadily [ˋstɛdəlɪ] (adv.) 穩定地
- ☑ recognize [ˋrɛkəgˏnaɪz] (v.) 認出；識別
- ☑ brand-new [ˋbrændˋnu] (adj.) 全新的

316

年終感謝與通知 E-mail 範例 1 中文

經銷商夥伴，您好，

　　首先，我要來對您表達我們的感謝之意，謝謝您們這一年來的努力！酵素科技公司持續穩定成長中，而有這樣的成果，多是因為有您們的協助，將酵素科技打造為一個全球知名品牌。在我們邁向新年度之際，我想要提醒您一件事，也跟您說說這幾個月內會發佈的一項令人振奮的改變。

假期出貨計畫
2014 年 12 月 24、25 日＋2015 年 1 月 1 日：不上班
我們在 2014 年 12 月 22 日、26 日、29 日這幾天會上班，但有特別安排才會出貨

新網站
我們在這下個月將推出全新的網站，也會在 20015 年初推出完全更新的型錄。

　　再次謝謝您們所有的努力，對於 2015 年、對於我們之間一直保有的夥伴關係，我們滿心期待！酵素科技的所有人員祝您們有個歡樂滿滿的佳節時節！
祝好
米凱爾・柏高拉
酵素科技董事長

Check 好句

☑ First of all, I want to express our gratitude for your hard work this year.　首先，我要表達我們的感謝之意，謝謝您們這一年來的努力。

☑ All of us at EnzymeTech wish you a very happy and healthy holiday season!　酵素科技的所有人員祝您們有個歡樂滿滿的佳節時節！

年終感謝通知 E-mail 範例 2 · 回顧展覽、通知放假日、祝福

Dear partner of MED Device,

It is hard to believe that 2014 is nearly over and that we start to look towards 2015! We'd like to take this opportunity to give you a brief review of MEDICA and some more information for 2015:

MEDICA: Thanks to those of you who came for visiting our booth at the MEDICA in Düsseldorf. The show was very successful for us and we hope it was for you as well.

LEGAL HOLIDAYS: We will be closed on 24th and on 31st December, 2014 and 1st January, 2015 due to legal holidays. The last Friday of 2014 for international shipping will be 19th December, 2014.

SEASON'S GREETING: It has been a prosperous year for us, a success we could not have accomplished without the trust of our customers. We appreciate your relying on our products and services over the past year. We'd like to send our best wishes to you all for a delightful Christmas season and a Happy New Year 2015!

All the best and kind regards,

MED Device Corp.

單字 ShabuShabu 一小補小補！
- ☑ device [dɪ`vaɪs] (n.)
  設備；裝置
- ☑ legal [`ligḷ] (adj.) 法定的
- ☑ greeting [`gritɪŋ] (n.)
  賀詞；問候語

年終感謝通知 E-mail 範例 2 中文

美迪醫療設備公司的夥伴，您好，

難以相信 2014 年就快要結束，我們也準備著要迎接 2015 年了！藉此機會，我們想跟您簡短回顧一下 MEDICA 醫療展，並告訴您 2015 年的一些訊息：

MEDICA 醫療展：謝謝在杜賽道夫 MEDIA 醫療展中前來我們攤位參觀的人，這次的展覽對我們來說非常成功，希望對您及／或您的公司也是同樣成功。

國定假日：2014 年 12 月 24 日、31 日，以及 2015 年 1 月 1 日為國定假日，2014 年最後一個星期五的出口出貨日，將會是在 2014 年 12 月 19 日。

佳節祝福：過去這一年對我們來說是個豐收年，若是沒有客戶的信任，我們也不可能達到這樣的成績。感謝您在這過去這一年來，如此信賴我們的產品與服務，在此獻上我們最誠摯的祝福，希望您們有個愉快的聖誕佳節，2015 年快樂一整年！

祝一切順利！
美迪醫療設備公司

Check 好句

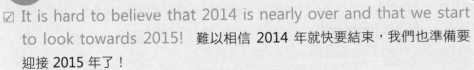

☑ It is hard to believe that 2014 is nearly over and that we start to look towards 2015!　難以相信 2014 年就快要結束，我們也準備要迎接 2015 年了！

☑ It has been a prosperous year for us, a success we could not have accomplished without the trust of our customers.　過去這一年對我們來說是個豐收年，若是沒有客戶的信任，我們也不可能有這樣的成績。

国贸英语 溝通術
Master English Communication for International Trade

電話對話

電話討論年度表現 範例

人物介紹

Michael

邁克，經銷商總經理，擅長分析市場，也擅長表現得令人激賞！

Alice

愛麗絲，原廠國際業務部經理，行動派，資訊抓得快！

Michael: Hello. Michael speaking. How may I help you?

Alice: Hi, Michael. This is Alice from MED Device. Happy New Year!

Michael: Hi, Alice. Happy New Year! Nice to hear your voice! How've you been?

Alice: Great! Especially after I saw your sales performance of 2014!

Michael: We're doing quite good, right?

Alice: You bet! We're impressed that you not only achieved the set growth target, but did even better!

Michael: We did invest lots of time and put considerable resources into managing existing customers as well as developing new customers.

Alice: We're very interested in knowing more about the challenges and opportunity you're facing in your market. I'd like to schedule a conference call for you to share your experience with us. Okay?

Michael: Sure! I would also love to hear your thoughts on our market!

Alice: Not a problem!

Michael: Great! I look forward to it!

PART 1

PART 2

PART 3

PART 4

PART 5

PART 6

PART 7

電話討論年度表現 範例 中文

邁　克：哈囉，我是邁克，請問有什麼需要我效勞的地方嗎？

愛麗絲：嗨，邁克，我是美迪醫療設備的愛麗絲，新年快樂！

邁　克：嗨，愛麗絲，新年快樂！很高興聽到妳的聲音！最近好嗎？

愛麗絲：很好！尤其在我看了你公司 2014 年的業績表現之後呢！

邁　克：我們做得還不錯，是吧？

愛麗絲：沒錯！你們不只達到設定的成長率目標，還超標呢！

邁　克：在現有客戶的經營上，和新客戶的開發上，我們的確投注了許多的時間，也投入了相當多的資源。

愛麗絲：我們很想多多了解你們在市場上所面臨到的挑戰與機會，我想安排個電話會議，請你跟我們分享一下你們的經驗，好嗎？

邁　克：當然可以！我也想聽聽你們對我們市場的想法！

愛麗絲：沒問題！

邁　克：太好了！我非常期待！

電話英文短句—好說、說好、說得好！

・Nice to hear your voice!　很高興聽到你的聲音！

・How've you been?　最近好嗎？

・I look forward to it!　我非常期待！

國貿知識補給站　併購大代誌

組織變革一向是件大事，若遇到的是併購事，那代誌就真的大條了！這年頭，國外企業大者恆大，愈變愈大，愈大就又愈併愈多！併購事對併購與被併購雙方的衝擊一定大，而對這些涉入之國外原廠的在台代理商而言，也是變天等級的大變化啊！

當你收到併購公告或通知時，你通常會看到上頭寫說一切基本上會是外甥打燈籠一照舅！但是光照舅舅還不夠，之後總是會再照到別的人，總是會有變化！

若你是併購廠商這一方的代理商，那麼，併購消息可是會讓你雀躍不已，因為那代表著你代理的廠家能做的餅更大了，產品種類更多了，若是不意外，被併購一方的產品應該也就併入你的代理範圍裡。不過，每一家廠商都已有其原本的代理商與通路，被併購一方的代理商一定會找舅舅（我知道夠了，我等一下不會再提舅舅了），要求一切作業如舊，所持的邏輯與理由充足之至，像是：

➤ 我們已多年深耕、經營被併購廠家的產品⋯⋯

We've been developing and dealing with the market for years⋯

➤ 我們與許多客戶已建立了良好的關係⋯⋯

We've worked with a wide range of clients and established good relations with them⋯

➤ 我們這麼多年來在市場開發上與在您品牌形象的建立上投資那麼多⋯⋯

We've invested so much in market development and in building your brand image⋯

322

　　這些是極為正當、極需發聲、極需大力爭取的代理權益,若是併購廠家考慮到確實被併購廠家的產品有其特殊性與差別性,那麼在其整合規畫上,他們就會採取先照舊運作的策略(我沒提舅舅喔),留住原代理,讓其繼續銷售被併購廠家的產品。

　　讓我們也來看看併購一方的代理商會怎麼說,他們也可提出邏輯與理由充足之至的論點如下:

> 我們的客戶要求要透過我們來購買被併購廠家的產品……
>
> Our customers want to buy from us the merged company's products...

> 我們對被併購廠家的產品與市場也很熟悉……
>
> We're also familiar with the merged company's products and markets...

> 我們可配置資源,專注開發該市場……
>
> We can allocate specific resources to develop the market…

　　你看出來了嗎?國外原廠併購消息一出,就是各海外代理公司角力、論述力與 e-mail 力競賽的開始。有一種繞口令的情況會讓被併購廠家的原代理所提的各種理由自動退下,結局立現,那就是,「當被併購廠家的代理同時也是併購廠家主要競爭廠家的代理時」,糊了嗎?簡單來說,若你這個代理商也代理了敵人的廠牌,那麼我就不給你理,不理你了!

　　在併購後的變革處理作業上,大事小事都有,從新訂銷售目標與策

略、客戶拜訪與說明,到行政上的型錄重印、訂單下單新方式、產品新型號、文件新格式、匯款新帳號等等,許多事都得調整調整。併購整合總得要經過大小調整不斷的過渡時期,在作業上實作的人,請記得在作業前不嫌麻煩地將新作法確認清楚,這樣才不會浪費心力在修正錯誤上,也才能確實地接上快速行進的整合步調!

《Dear Amy》時間　大哉問：要怎麼樣才能學好英文？

　　我們每個人幾乎都問過這個問題，如果你因為害羞而不敢開口問別人這個問題，那麼你心裡一定想這個問題想過上百遍了。我的一個英文很好的朋友問一個英文更好的朋友：「怎樣才能學好英文？」這位英文好得很的朋友回了這樣的話：「看你要做什麼用呢？」，這個反問這就是答案，就是道理，就是真理啊！「在哪裡？」若你還沒參透其中深沉的意涵，那我就要來問你：「那你要到哪裡？」…好了好了，我就不追著這裡那裡說了，我的意思是說，在英文學習上，你要先想清楚你學英文是要做什麼用？若是要讓自己出國旅遊時可看得懂標示，問得了路，說得通，那就請妳鎖定旅遊相關的英文資料，多訓練聽與說。若是你想要增加自己英文閱讀的能力，那就請問問你自己，你想要讀懂的是什麼樣的內容？小說嗎？那容易，就請找一本你翻看時看得懂七成左右的小說來看著學習，看著享受。之所以要說抓個七成懂的書來看，是因為這樣的書才能讓你有一直讀下去的興趣，「那我就找一本不太懂的書，邊看邊查，這樣我不是可以學得更多嗎？」若你肯下這種「苦修」的功夫，那當然好，但請注意不要一本苦讀個兩年都還沒讀完啊！若是你最想要搞懂工作上會用的英文，可用在工作上，那很好，因為這樣的目標是明確的，範圍基本上來說是特定的。請你找一本商用英文學習書…沒別的好說了，就這一本啦！請你「從頭看到尾」，為什麼要強調這點？請你回頭看一下你的書櫃，是不是有好多本英文學習的書，啊好多本都沒看完？是的！只要你每一本都好好看，你的英文程度多少都一定會提升！除了看商用英文的書，當然不要放過最適用、最實用的英文素材——那就是你工作上接觸到的任何文件與資料上的英文，從品名、型錄、產品技術文件、文宣，到國外原廠發來的通知信函、e-mail，以及原廠的網站內容。當你認命、願意又樂意地開始循序漸進用心專心學英文，那不久後你就一定可以找到對英文的信心了！有了專心，又有了信心，世上哪還有什麼難事呢！

單字片語說分明

- merger [mɝdʒɚ]

n. （公司或機構的）合併 the process of combining two companies or organizations to form a bigger one

例 What are the benefits for customers from the merger?

這個合併對客戶來說有什麼好處呢？

merge [mɝdʒ] v 使合併；融合

常見搭配詞

merge into/ with 　　　　　　　　merge into the background

融合；融入 　　　　　　　　　　　不引人注意

- acquisition [͵ækwəˋzɪʃən]

n. （公司的）收購 a company that is bought by another company

例 The official announcement regarding the acquisition will be made next week.

關於這次收購的正式宣告，將會在下個星期發佈。

其他字義：取得；（技能或知識的）習得；購得物

acquire [əˋkwaɪr] v 取得；獲得

acquisitive [əˋkwɪzətɪv] adj 渴望得到的；迫切求取的；貪婪的

常見搭配詞

newly acquired 　　　　　　　　acquire a taste for sth

剛取得的 　　　　　　　　　　　開始喜歡某事物

acquired taste 　　　　　　　　language acquisition

養成的嗜好 　　　　　　　　　　語言習得

- takeover [ˋtek͵ovɚ]

n. 併購；接管 a situation in which one company takes control of another company by buying a majority of its shares

例 What will happen to us if this takeover really occurs?

如果併購真的發生了，會對我們造成什麼影響呢？

take over 接手；接替；接任；接管

常見搭配詞

| a takeover bid | a hostile takeover | a military takeover |
| 併購出價 | 惡意併購 | 軍事接管 |

· alliance [ə`laɪəns]

n. 結盟；聯盟 an arrangement between two or more people, groups, or countries by which they agree to work together to achieve something

例 These two conglomerates have announced that they are to form a strategic alliance.

這兩大集團已宣布將組成策略聯盟。

ally [ə`laɪ] Ⓥ 與…結盟；與…聯合

常見搭配詞

enter into / form / forge an alliance 結成同盟

in alliance with sb 與某人結盟

· affiliate [ə`fɪlɪ͵et]

v. 與…（更大組織或團體）有緊密關聯 to have a connection with or support a larger organization or group

例 Please contact our affiliated company in your region with the contact information below.

請您與我們在您區域的分公司連絡，連絡資料如下。

其他字義：隸屬於某事物

affiliate [ə`fɪlɪɪt] Ⓝ 分公司；附屬機構

ally [ə`laɪ] Ⓥ 與…結盟；與…聯合

- subsidiary [səb`sɪdɪˌɛrɪ]

  n. 子公司 a company that is owned by a larger company

  例 Our Group has more than 10 subsidiaries in different industries.

  我們集團有十幾間子公司，跨足不同的產業。

  subsidiary [səb`sɪdɪˌɛrɪ] adj. 附帶的；附屬的；次要的

- capability [ˌkepə`bɪlətɪ]

  n. 能力；才能 the ability to do something

  例 Our company has the capability and skills to develop design concepts into products.

  我們公司有能力，也有技術，能將設計概念轉化成產品。

  其他字義：性能；（國家的）戰鬥力

  capable [`kepəbl] adj. 有能力的；有才能的；能幹的

  常見搭配詞

  within sb./ sth.'s capability   beyond sb./ sth.'s capability
  在…的能力範圍裡   超出…的能力範圍

  military/defense capability   nuclear capability
  軍事能力   核武能力

- synergy [`sɪnədʒɪ]

  n. 合力；協力 the extra energy or effectiveness that people or businesses create when they combine their efforts

  例 The challenge for the Manager is to create synergy among these very different groups.

  這位經理面臨的挑戰在於要如何從差異甚大的幾個團隊中創造出綜效。

- streamline [`strim,laɪn]

  v. 使現代化；使簡化；使有效率 to improve a business, organization, process etc. by making it more modern or simple

例 We are committed to helping you streamline the experimental process by offering you our best and most up-to-date equipment.
透過我們所提供的最好且最新的設備，我們保證能夠幫您簡化實驗流程。
streamline [`strim͵laɪn] n 流線型 adj 流線型的

· diversify [daɪ`vɝsə͵faɪ]
v. 使多樣化；擴大經營範圍 to develop new products or activities in addition to the ones that you already provide or do
例 We can help you to diversify your investment in an efficient and cost-effective way.
我們能以有效率、具成本效益的方法，幫助您讓投資多元化。
diversified [daɪ`vɝsə͵faɪd] 多變化的；各種的
diverse [daɪ`vɝs] adj 不同的；多變化的
diversification [daɪ͵vɝsəfə`keʃən] n 多樣化；多角化經營
diversion [daɪ`vɝʒən] n 轉移；分散注意力；分散注意力的事物；消遣

· prosperity [prɑs`pɛrətɪ]
n. 繁榮；昌盛；富庶；成功 the situation of being successful and having a lot of money
例 We wish you a happy holiday season filled with joy, peace and prosperity!
我們祝您在這歡樂佳節的季節裡，擁有滿懷的歡欣、平靜，還帶有豐收的希望！
prosperous [`prɑspərəs] adj 繁榮昌盛的；富足的；成功的

- **accomplishment** [ə`kɑmplɪʃmənt]

  n. 成就 something difficult that you succeed in doing, especially after working hard over a period of time

  例 From a technical standpoint, it's really a remarkable accomplishment.

  就技術面來看，這真是一個重大的成就。

  accomplish [ə`kɑmplɪʃ] ⓥ 達到；完成

  accomplished [ə`kɑmplɪʃt] adj 完成了的；已實現的；熟練的；有造詣的

- **gratitude** [`grætə͵tjud]

  n. 感激；感謝 a feeling of being grateful to someone because they have given you something or have done something for you

  例 Words cannot adequately express our gratitude to you!

  我們對您的感激之情，溢於言表！

  grateful [`gretfəl] adj 感激的

好書報報

Leader 013
國貿英語溝通術

作 者	劉美慧	
發 行 人	周瑞德	
企劃編輯	徐瑞璞	
執行編輯	饒美君	
封面設計	高鍾琪	
內文排版	菩薩蠻數位文化有限公司	
校 對	陳欣慧 陳韋佑	

印 製 大亞彩色印刷製版股份有限公司
初 版 2015 年 2 月
出 版 力得文化
電 話 (02) 2351-2007
傳 真 (02) 2351-0887
地 址 100 台北市中正區福州街 1 號 10 樓之 2
E - m a i l best.books.service@gmail.com
定 價 新台幣 360 元

港澳地區總經銷 泛華發行代理有限公司
地 址 香港新界將軍澳工業邨駿昌街 7 號 2 樓
電 話 (852) 2798-2323
傳 真 (852) 2796-5471

國家圖書館出版品預行編目(CIP)資料

國貿英語溝通術 / 劉美慧著.-- 初版.-- 臺北市
：力得文化, 2015.02
面 ； 公分. -- (Leader ; 13)
ISBN 978-986-91458-1-7(平裝)

1.商業書信 2.商業英文 3.商業應用文 4.電子郵件

493.6   104000998

力得文化
Leader Culture

Lead your way. Be your own leader!

力得文化
Leader Culture

Lead your way. Be your own leader!